CLEAN TECHNOLOGY

CLEAN
TECHNOLOGY

ALLAN JOHANSSON

LEWIS PUBLISHERS
Boca Raton   Ann Arbor   London   Tokyo

Library of Congress Cataloging-in-Publication Data

Johansson, Allan.
    Clean technology / Allan Johansson.
        p. cm.
    Includes bibliographical references and index.
    ISBN 0-87371-503-9
    1. Economic development — Environmental aspects. 2. Industry
— Environmental aspects. 3. Environmental engineering. I. Title.
TD195.E25J64   1992
333.7—dc20   91-46683
                                                CIP

Direct all inquiries to CRC Press, Inc., 2000 Corporate Blvd., N.W., Boca Raton, Florida 33431.

Printed in Mexico   1 2 3 4 5 6 7 8 9

Printed on acid-free paper

Author

Allan Johansson, born in Finland, received his degree in Chemical Engineering, and after studies in the Theoretical Chemistry Department of Cambridge University, U.K., he received his Doctor of Technology degree from the Helsinki University of Technology in 1973.

After a period of time as a Senior Consultant for the pulp and paper industry and later at the Battelle Geneva Research Centres as leader of the process development section, he joined the Technical Research Centre of Finland (VTT) as a research professor, where he now directs the Non-waste Technology Research Unit which is focusing on technical development aiming at reducing wastes and other adverse effects of industrial activity.

Foreword

The last decades have seen a shift of emphasis in industrial policy from technical issues towards social issues such as improved working conditions, public safety and, lately, environmental concern, extending the responsibility of engineers beyond the traditional realms of technology. We have to realize, however, that the present environmental threats as manifested through alarming changes in the biosphere are a result of the magnitude of today's industrial operation rather than a result of specific technological shortcomings. Most of the questions concerning the direction of future development of mankind are in fact not technical at all. Some guidelines do, however, emerge from the aspirations towards sustainable development that are of relevance for today's and tomorrow's engineers.

These are perhaps best described by the expression "Clean Technology", and above all involve an effort to minimize effluents and waste production, but also the design of products as well as the possibilities to offer the same services to the consumer, from the ecological point of view in an entirely different, more benign way.

Although many of these issues are dealt with in the specialized literature and at conferences, it is difficult to find them in a collected form suitable for giving an introductory overview of the changing situation, as well as the technical possibilities dealing with emerging problems. It is the author's experience that a broad understanding of different technologies is very helpful in developing new process concepts.

This text is an effort to give a broad introduction into the field of Clean Technology, which in its details, is not such a new concept after all, but rather an application of plain "good old engineering" where some of the distorting economic constraints have been replaced by more recent environmental and resource management considerations. For this purpose, technologies have been included in the text which are not yet commercial but may well find use in a particular context due to their performance.

As certainly many others have done in similar cases, I have come to realize that the undertaking of an effort of this kind, although noble in its ambition, is nearly impossible to fulfill in practice with satisfaction, and the final result must always be a compromise between the necessary and the possible. It also requires a larger than usual tolerance from one's surroundings. I therefore express my deepest gratitude to all those that have supported the task, notably the Technical Research Centre of Finland whose staff has continuously given their support to the cause of combining technical progress with environmental consideration without which this work would have been impossible.

Allan Johansson
Helsinki
February 1992

Contents

CHAPTER 1

Introduction

Little did the prehistoric man or gorilla care about physics when he grabbed a branch to add strength to a blow, perhaps the first forceful demonstration of the powers of technology as a means of fighting nature.

Tradition lingers on; even today, one could say that discussions concerning technology and the environment reflect an almost mythical conflict. Technology and technologists are often considered as some kind of antienvironmentalist. This labeling is by no means accidental or even unjustified, as the very objective of technical development until recently has been to fight against nature and to help conquer the earth for the human being.

This mastery of the earth, fully realized only in this century, has transformed man from the object of evolution to its subject. The resulting worldwide accountability, extending well into the future, is an even greater event in the history of earth that was the first appearance of humans several millions of years ago. Only 20 generations ago nature still was a threat to mankind — an overwhelming adversary. Only a few generations ago every cleaning of land, every founding of a new city, was celebrated as a victory in the battle against the inhumane adversary "Nature", the "wilderness". Science and technology freed us from the yoke of nature. The monstrous giant was overcome, domesticated, and put out to pasture. In view of this most recent reversal in the meaning of nature, it is perhaps easier to understand why Homo sapiens sees his role with regard to nature as that of a plunderer rather than that of a caretaker, why we continue to view nature as a resource to be exploited, why we pervert our need for security into a threat of total annihilation, and why our need to consume allows other human beings to be degraded into little more than machines.[1]

Unfortunately, today we can say without too much satisfaction that both engineers and scientists have been an efficient lot; the objective has been attained perhaps too well. The ancient dream of Francis Bacon envisaged in his "Dominion of Man over the Universe" has finally come true; footprints of technology can be seen almost everywhere on earth.

In view of this, one might ask whether the present ecological threats are nature's way of striking back, the "nemesis" as an answer to man's technological "hybris" that the contemporary philosopher Georg Henrik von Wright[2] warned us about. Regardless, the fact remains that from an early and relatively modest beginning, we have come a long way. Massive use of technology resulted in what we now call the industrial era, a period of little more than a century of rapid and accelerating growth of industrial production and wealth, and — as we today are well aware — environmental problems. However, although the discussion about the negative effects of technology on the environment is a relatively recent one, the problem itself is not new. An early example of the dramatic effects of human activity upon its surroundings is perhaps the almost complete deforestation of the Greek archipelago some 2000 years ago.

What is new is that everything today happens on a grander scale; instead of local problems, we are now facing global threats as a result of too intensive industrial activity.

Credit for the first serious cry of warning generally goes to Rachel Carson[3] for her book *Silent Spring* describing the harmful effects of DDT on the ecosystem. At that time, the new and alarming issue was the discovery that harmful effects or chemicals could accumulate over a long period of time and cause problems far from the actual source. Perhaps an even greater surprise was the vigorous response of the general public, which up to then had shown little or no interest in global environmental threats.

With that, a new era of enlightenment began. The public requested and got more information about what was happening in our environment, or, rather, what some people thought was happening or could happen. The great number of different opinions, the factual basis of which were in many cases meager, to put it mildly, initially resulted in an atmosphere of confused despair and, later, hostility against technological development and what was commonly called "The Techno-economic Establishment".

Thus, it is not surprising that the voices calling for technical alternatives and changes in the direction of technological development have grown louder, sustainable economic development being the ultimate test. While we are waiting for the perfect solution to emerge, a rational technological approach would be to strive toward a minimization of effluents — all the effluents produced by industrial activity. But the problem does not end there. In fact, as we shall later see, industrial effluents constitute only one side, although a very visible one, of a multifaceted problem: the role of technology in tomorrow's society.

REFERENCES

1. Neuweiler, G. *Evolution und Verantwortung,* Alexander von Humbolt Stiftung Mitteilungen Hamburg, 1986, 48, 1.
2. von Wright, G. H. *Vetenskapen och Förnuftet,* Söderstöm & Co., Förlags Ab, Helsingfors, 1986.
3. Carson, R. *Silent Spring,* Houghton Mifflin, Boston, 1962.

CHAPTER 2

Background

2.1.   INDUSTRIAL SOCIETY

In terms of the time scale of biological evolution the history of Homo Faber, that of the ingenious industrious man, is very short indeed. Of this episode, the industrial era in which we are presently living is characterized by an enormous, almost explosive expansion of the production machinery.

This expansion was made possible by scientific and technical progress, but also by the existence and universal acceptance of an almost limitless market of goods and services. In the beginning of this era, these needs were real and in all respects morally justifiable (a situation which does not always hold true anymore). Thus, it is not surprising that during the early period of rapid expansion of the industrial society, nobody could actually believe or anticipate the problems that would later be encountered.

As far as the "almost limitless market of real and morally justifiable needs" goes, this essentially still holds true. Today, four fifths of the earth's population still lacks industrial wealth; however, the terms of development are not dictated by the needs of this majority, but by the wishes of the more fortunate one fifth. The situation might be easier to resolve if this were not the case.

We are living in a world of enormous polarization of wealth and power not necessarily founded on cultural superiority (or other mental dimensions of equal importance for well-being) in which the ecological problems have assumed proportions that threaten our existence as a biological species.

The following briefly sketches the history of the evolution of the industrial society from the point of view of technical development.

In general, the liberation of man from the limitations imposed on him by his modest muscular power, the beginning of the industrial era, marks the first *turning point!* (Actually, one could go a bit further and consider the discovery of fire some 400,000 years ago as the real beginning of man's liberation from the constraints imposed upon him by nature.)

In the beginning of industrialization, many values were carried over by tradition — the notion of quality was still important as a tradition from manual handicraft.

As mentioned earlier, there were many needs to fulfill, a limitless market, and very little competition. From this situation evolved one singularly important guideline — maximize production! This maxim conditioned much of the technical development and was the origin of many of the side effects we suffer from today; e.g., inefficient raw material use and maximization of wastes in production (but not yet in society).

Along these lines, a very efficient production machinery was created, eventually resulting in market saturation and widespread and intense competition. In this new situation, two options presented themselves: compete with superior quality of product or with technical ingenuity and price. The former option represented the traditional conservative thinking, the latter the new progressive attitude. Both routes were exploited; in general, the latter became predominant, as it often was faster and carried in itself the seed for creating new needs. Technology not only filled existing needs, but created new ones, real or imagined! This marks the second important *turning point* in the evolution of the industrial society.With this new aspect, the time factor in production became even more important. It became important to produce and sell as much as possible as long as the product was competitive — a mad race in which it has become more and more difficult to recover the ever-escalating costs of research and development.

Today, our industrial society faces many serious problems, such as (1) market limitations, (2) raw material problems, notably for energy production, (3) the enormous energy dependence which is a real threat, and finally, a menace of new dimensions, (4) global environmental pollution.

Never before in history have the waste streams of man's industrial activity reached such proportions that they could threaten the very existence of the human beings on earth. Environmental problems on a global scale mark the third *turning point* for Homo Faber, whose present-day descendant should perhaps more properly be called Homo Profusius, the wasteful man.

Not even the most convinced technocrat could, in the beginning of the industrial era, have predicted that the growth in industrial production and wealth would take such gigantic proportions as has happened during the last few decades. Hence, such questions as possible overexploitation of the earth's resources or environmental pollution were largely ignored in the beginning of industrialization — and remained so for too long, it appears.

Today, although only one fifth of the earth's population enjoys the fruits of industrialization, we are already seeing the limitations of growth, and some of the fundamental paradigms that have so success-fully contributed to the industrial revolution must be questioned.

The greatest achievement of the coming decade in this area will most likely be the fundamental change in attitudes toward a more consistent environmental policy of industrialists and governments alike. Concerns such as environmental policies and waste management are today discussed openly between industry and governmental authorities in sincere efforts to find practical solutions to our dilemma, in a manner which would have been unthinkable only 10 or 20 years ago.

Progress is made, but only talk is not enough! Presently, our knowledge about environmental threats seems to grow faster than our ability to cure them. Also new is the fact that many of the consequences of overexploitation of the earth are resulting in global effects that threaten our future rather than, as previously was the case, in local problems of limited spatial and chronological consequence.

It must also be said that our present technology for dealing with waste and environmental problems is remarkably unsophisticated. We are only at the beginning of an environmentally oriented technology development. Personally, I am convinced that in the next decades, environmental considerations will be one of the major driving forces in the development of new technology, and I see here an enormous and as yet rather unexploited reserve for putting development on the right track again.

The problem, of course, is that we are rapidly running out of time. In a manner of speaking, we are to a great extent prisoners of our past. We have created an enormously efficient production machinery which in its present scale of operation no longer fits into the ecosystem, but for which we at present have no valid substitute. And even if we had, it would take time to change the course, time we may not have.

We are speaking of fundamental changes in the whole industrial society. Two conceptual possibilities exist:

1. We create, in a thermodynamic sense, a closed industrial system which exchanges only energy (no material) with nature, and all material flows are confined to the system. Product flows return and are reprocessed in a truly recyclable manner.

2. We develop an industrial production system totally compatible with nature, "soft" technology using renewable raw materials and biodegradable products.

In the first case, we have to decide what should be recycled. Traditionally, one tends to think of recycling only as a method of reutilizing the raw material content in the production process. This already is current practice within the metallurgy sector, and quite frequent within the pulp and paper industry as well as the glass industry. In all these cases, it makes sense, and in addition money, without elaborate changes in the production cycle. Simple common sense and good old-fashioned engineering is required.

A less appreciated, but potentially even more rewarding recycling strategy would be to recycle products. In that way, not only the material content of the product would be reused but also much of its labor and design value. In a manner of speaking, we would also recirculate part of the entropy content of the product. In fact, although perhaps less well known, this type of activity already exists on an industrial scale in certain sectors, notably the automobile industry and some sectors of the tool and appliance sector.

This is certainly an area that could be developed much further by incorporating the strategy at the design stage of different products. It would also have some far-reaching and positive consequences on aspects such as product quality. The longer the planned lifetime of the product, the more it is worth putting into it.

To a certain extent, one can see signs of this approach gaining ground, and it is interesting to note that contrary to what was once believed, high-quality products such as cars and cameras, even within the consumer sector, never totally lost ground to the modern, short-lived high-technology versions.

An unfortunate side effect of more sophisticated technology and products is the increased use of unconventional materials which often, in their particular behavior, include toxicity, even in small quantities. It is also typical of high-technology products that their design and manufacture value is much higher than the value of their material content. During their useful life, the value is represented by order, not quantity. Although expensive raw materials are frequently used in these components, they are used in minute quantities and, in fact, dispersed through these products when they become wastes.

Recycling, although potentially appealing, seems to be a very inefficient strategy when dealing with such wastes. In order to be efficient, one would need an absolutely foolproof collection and recycling system, as only a minute fraction of the waste stream can create serious hazards if left unattended or disposed of improperly. Either these products would have to be banned, which is difficult as the hazards are often known only later or when the damage is already on its way, or we must devise a waste disposal system which includes a final and environmentally acceptable destruction of the "tailings", the inevitable final waste of even the wasteless society. Probably both alternatives must be considered.

The second alternative, soft technology, involves a more dramatic re-evolution of our industrial society. All the recycling alternatives of the first approach would still be valid and make sense within this imaginative world, with the difference that the final residues, being biocompatible, could be taken care of by nature. Although this would appear to be an ideal situation worth striving for, it is difficult to believe that such a utopia could ever be achieved without serious cuts in our current living standard. The early farming society satisfied, to a very large extent, the criteria for total biocompatibility, but it would, of course, be hopelessly inadequate for satisfying our present-day needs (see Table 2, Chapter 3).

On a more fundamental level, one might ask if biological processes are not, in general, too slow to satisfy the production rates of a modern industrial society. Both approaches thus have their inherent drawbacks, and what is likely to emerge is a hybrid strategy between the two. One should take advantage of the efficiency of our present production knowledge combined with biocompatible products in order to, on the one hand, satisfy the enormous production volumes required and, on the other hand, avoid the insurmountable difficulties linked to the requirement of total recycling. Also, the notion of good quality products with long lifetimes, worth the maintenance needed by high-technology products, deserves attention.

The most direct method for changing the situation for the better lies in changing consumer habits (choice of products) as well as the willingness to handle wastes in a more conscientious manner. Such a change requires no elaborate structural or legal changes; it does, however, require correct and continuous information.

There is a real need for a change in the traditional approach. We cannot continue to remain only passive problem solvers. We must look ahead and formulate strategies for sustainable development, take a more responsible and more holistic view. In a manner of speaking, we are about to graduate to a higher level of industrialization, rather than speak of science and technology, one should perhaps introduce a new concept — scientific technology.

2.2.   RESOURCE LIMITATIONS

The guiding principles for the evolution of species in nature has been optimization of resource management (energy vs. food), which has led to a great variety of different solutions adapted to local conditions. Ectotherms, for example, do not generate their own body heat and thus generally require less food (energy) per unit of body weight than endotherms. For example the resting metabolic rate for a 20-g lizard is about 1 cal/g/h, whereas a 20-g mammal consumes nearly ten times more. As a result, ectotherms are able to survive in areas where the energy flow to the ecosystem is small or unpredictable, since they use less energy while waiting for the next source of food. The greater percentage of assimilated food available for growth and reproduction in ectotherms allows them greater efficiency in converting food into new biomass. The net efficiency in turning food into living tissue for mammals and birds ranges between 0.5 and 2.3% while that for amphibians and reptiles is 6.3 to 49%.[1] On the other hand, the higher and more constant temperatures of endotherms give them the possibility of a higher and more sustained range of muscular activity over a wider temperature range, with consequent ability to gather food where it is abundant. In this sense, endothermy is an energy investment that allows the large capture of available energy flows in the organized environment. According to what we know, neither evolutionary strategy is better, since both types obviously exist, depending on what is favored in specific environments or at specific times. [1] Also, migration is a question of the energy invested in migration for the benefit of a better energy-harvesting opportunity, by reaching regions of dense food, for example, or improved reproduction regions. [2]

Optimal resource management, in a broad sense, is also a central issue for the future of mankind. The oil crisis in the 1970s signaled a major threat to the industrialized world. Not only did the environmental pollution caused by industrial operation appear to be a threat to civilized life on earth, but people suddenly were told, in rather dramatic terms, that we were running out of energy, the very basis for the industrial expansion.

A closer look showed that the fossil fuel supply, or more specifically, the oil supply, was not the only necessity that was rapidly running dry at the present rate of consumption; many other vital raw material resources turned out to be surprisingly small or unevenly distributed geographically in relation to their consumption rates. The latter is of obvious political and economic importance.

Credit for the first attempt to make a comprehensive assessment of the alarming situation is generally given to a group of concerned citizens better known as "The Club of Rome". In their publication, *The Limits to Growth,*[3] they outline different scenarios for world development. Today, we can, with the wisdom of hindsight, note that the present situation looks far better than even the most optimistic scenario of that time, no doubt partly due to the enormous impact of the predictions and corrective actions taken at the time.

Personally, I believe that the first oil crisis was really a blessing to the industrialized world, in the sense that it brought us down to earth again, and caused us to face at least some of the real problems around us. It is also comforting to realize, I think, how much improvement could be achieved by technical development within the energy sector once the priorities were set.

We are all aware of the central importance of energy for industrial production. In general, there seems to be a direct correlation between the per capita energy consumption and the per capita income of a country and we shall address this matter in considerable detail in the next chapters. But first, a few words about resources in general.

2.2.1. Forests

Tropical forests extend over 20% of the land on earth, but are diminishing rapidly: closed forests at the rate of 7.5 million ha/year and open forests at 3.8 million ha/year. According to one estimate,[4] at least an additional 225 million ha of tropical forest, out of a total of 900 million ha, will be cleared or degraded by the end of the century. This is a change of far-reaching consequences as forests are believed to shelter the great majority of the planet's species. Lack of genetic diversity may result in a fatal lack of resistivity in a changing world. It has been estimated[5] that we are currently experiencing a loss of 7000 species per year, which is about 10^3 to 10^4 times the natural rate of extermination.

2.2.2. Water

The total volume of water on earth is about 1.6 billion km^3, some 10^{18} t. More than 97% of this is saline seawater; of the rest, 22% is groundwater and 77% is ice locked in the glaciers and the polar ice caps.[4] Only about 0.014% is available to support earth-bound creatures. Even this modest proportion would ensure an adequate supply of freshwater now and for the foreseeable future were it not, as most resources usually are, located where it is least needed (Figure 1).

Water shortage has consequently been one of the most critical issues in many arid regions (Table 1). Today, this problem is not limited only

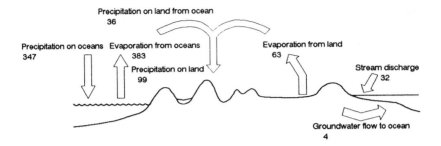

Figure 1. Global water cycle (10^3 km^3).

to arid regions. Scarcity and heavy pollution have resulted in the fact that at least one fifth of the people living in cities in the third world and three quarters of its rural people lack access to reasonably safe supplies of water. Intensive agriculture is by far the greatest user of water, consuming about 70% of the world's use of fresh water (of which roughly two thirds is wasted), and it is also the greatest polluter. An alarming feature is that in many places, groundwater reserves already are showing signs of pollution.

2.2.3. Air

According to the Organisation for Economic Co-operation and Development (OECD),[6] in 1980, global emissions of common air pollutants into the atmosphere as a result of anthropogenic activity consisted of about 110 million t of SO_2, 59 million t of particulate matter, 69 million t of nitrogen oxides, 193 million t of carbon monoxide, and 57 million t of hydrocarbons (excluding methane).

Energy conversion and use is a major if not the most important, source of pollution. According to statistics of the Federal Republic of Germany in 1986, 99% of NO_x emissions, 96% of SO_2 emissions, 92% of CO_2 emissions, 88% of CO emissions, 55% of volatile organic compounds, and 43% of particulates originated from energy activities,[7] and the situation is similar in most industrialized countries.

The projected growth in energy demand implies the risk that the agreed upon reduction of pollution by the year 2010 may not be sufficient and may not result in the targeted levels unless additional control measures are taken.

Table 1
Water Reserves of the Biosphere

Oceans	1,370,000
Groundwater	320,000
Ice	16,500
Lakes	34
Humidity of atmosphere	10

Note: Reserves in 10^3 km^3.

In 1980, SO_2 emissions in Europe were 54.1 million tons, of which 54% was in Eastern Europe and 46% in Western Europe. According to government plans, SO_2 emissions would fall by 18% (Eastern Europe, 21%; Western Europe, 14%), to a level of 44.5 million tons in the year 2000 mostly achieved through more rational use of energy, fuel substitution, and the broad application of desulfurization techniques in the power generation sector.[8]

In the European Economic Community (EEC), the projected increase of energy consumption between 1985 and 2010 (210 million toe [ton oil equivalent] or 23%) may still be associated with a decrease of SO_2 emissions by 45% or 6.4 million tons, provided appropriate measures are taken.[8] An amendment to the Clean Air Act presented before the U.S. Congress is intended to reduce SO_2 emissions from power stations by half, or 10 million tons, by the year 2000.[8]

In the market economies, NO_x emissions are predicted to decrease in the longer term. In the European Community, NO_x emissions are likely to grow from 7.7 million tons in 1985 to 9.6 million tons in 2000, but fall thereafter to 9.3 million tons in 2010.[8] The proposed amendment to the U.S. Clean Air Act aims at a 10% reduction of NO_x emissions from power stations and 40% from cars.

In the centrally planned economies, NO_x emissions are likely to increase due to a significant increase of motorization. At present, there are only 45 cars per 1000 inhabitants in the U.S.S.R., whereas there are 540 in the U.S.A., 400 in Canada, and between 200 and 400 in West European countries.

The total amount of CO_2 emissions from fossil fuel combustion in the EEC region will apparently increase from 1.5 billion (10^9) tons in 1985 to 17.1 to 18.8 billion tons in 2010 if CO_2 emission controls are not widely applied. Together with the other greenhouse gases (methane, nitrogen oxides, CFC-11, and CFC-12), these emissions could induce significant changes in the climate.

The amendment to the U.S. Clean Air Act envisages an 18% reduction of CO emissions from cars (not already equipped with reduction devices) and a 2 to 5% reduction of volatile organic compounds.

2.2.4. Soil

It has been estimated[4] that against 1500 million ha of land currently used for crop production, nearly 2000 million ha have been lost in historical times. At present 5 to 7 million ha of cultivated land (0.3 to 0.5%) are being lost each year due to soil erosion. If the present trend continues, it seems that all programs for adding more land to food production may not even compensate for the areas lost as a result of soil degradation and competing land uses.

The scale of human operations is best illustrated by a few figures giving the order of magnitude. Human society handles a flow of materials of $3.3 \cdot 10^{10}$ tons/year compared to, e.g., natural flows such as sedimentation 0.6 to $1 \cdot 10^{10}$ tons/year or the materials moved by all rivers before man's intervention ($0.93 \cdot 10^{10}$ tons/year, a value which has subsequently more than doubled due to human intervention).[5] Today, we consume materials at a rate of 10^3 tons/s and water at a rate of 10^5 tons/s.

Another serious problem threatening the soil is presented by the thousands of hazardous waste dumps that are the legacy of decades of uncontrolled disposal of toxic byproducts. These must now be located and cleaned up at great expense.

2.2.5. Material Resources

We are using up the fossil fuel reserves of the earth at an astonishing speed, as shown in Table 2. These reserves, which were formed by accident, one could say, millions of years ago from biomass entrapped in oxygen-free zones, can be used only once and are then lost forever.

As we shall later see, energy production is vital for all industrial activity, which explains the great attention devoted to suitable primary energy sources in many of the discussions dealing with the future of mankind. Fossil energy or energy in general, however, is not the only resource in short supply.

Table 2
World Energy Consumption

Energy	1970	1980	1986
Coal	2134	2728	3196
Lignite	791	1004	1225
Oil	2275	2982	2787
Gas	581	808	924
Total	5781	7522	8132

Note: Consumption in millions of metric tons: figures rounded.

Source: *United Nations Industrial Statistics Yearbook,* UN Statistical Office Energy Statistics Yearbook, United Nations, New York, 1982, 1989.

The expanding industrial activity needs materials, and the mineral resources of the earth are ruthlessly exploited, as seen in Table 3. Different from energy, there are no theoretical obstacles for reusing these materials; nevertheless, surprisingly little is recycled today. Great quantities are dissipated into the ecosystem (Table 4), and hence not only lost, but also creating additional environmental problems.

2.3.   ENVIRONMENTAL PROBLEMS (LOCAL-GLOBAL)

We simply have to accept the fact that in all our ingenuity, we have not become a lot wiser when it comes to avoiding environmental hazards; instead, the list of potential threats has grown longer:

- Increase of CO_2 content of the atmosphere (greenhouse effect)
- Depletion of the ozone layer (increased UV-radiation intensity)
- Pollution of the oceans
- Deterioration of the groundwater quality
- Accumulation of harmful chemicals in animals and man

The items in this list have not been placed in any particular order of importance and in fact this is not very important, as a common feature of them seems to be that they all are catastrophic in the sense that should any one of them get out of control, the living conditions on earth could be seriously altered. Moreover, we know far too little about them and the factors behind them.

Table 3
World Ore Consumption

Ore	1970	1980	1986
Iron	731	973	915
Copper	21	27	29
Nickel	70	70	70
Bauxite	58	94	89
Lead	62	59	73
Zinc	124	169	139
Manganese	27	37	32
Chromium	2.7	3.1	4.1
Limestone	649	870	735
Total	1745	2302	2085

Note: Consumption in millions of metric tons: figures rounded.

Source: *United Nations Industrial Statistics Yearbook,* UN Statistical Office Energy Statistics Yearbook, United Nations, New York, 1982, 1989.

Table 4
Natural and Man-Made Fluxes of Metals Dissipated into the Oceans

Element	Natural	Man-made
Iron	25,000	319,000
Manganese	440	1,600
Copper	375	4,460
Zinc	370	3,930
Nickel	300	358
Lead	180	2,330
Molybdenum	13	57
Silver	5	7
Mercury	3	7
Tin	1.5	166
Antimony	1.3	40

Note: Figures in thousands of tons per year.

Source: Glasby, G. P., *Ambio,* 17(5), 330, 1988.

The most alarming thought, however, is that in addition to the many environmental threats we are aware of, there must be many equally menacing ones still waiting to be discovered. The sooner we identify them, the better, as here too applies the truth discovered by many doctors long ago: "a disease is easier to cure in the beginning, but difficult to diagnose, and when it finally breaks out in full vigour it is easy to diagnose but difficult to cure".

Lest the picture be too gloomy, it is perhaps worth pointing out that the use of technology does have its environmental merits as well; in some cases, the situation even appears to be reversed from what it used to be. For instance, deforestation, which has become a serious problem to many developing regions, is more often than not due to insufficient use of technology and hence local overexploitation of the forests.

2.4.   SUSTAINABLE DEVELOPMENT (TECHNICAL IMPLICATIONS)

The notion of sustainable development has been much debated ever since it was put forth by the World Commission on the Environment and Development. In their report, *Our Common Future*,[9] they define sustainability as "Sustainable development that meets the needs of the present without compromising the ability of future generations to meet their own needs".

There are two key issues:

1. needs, in particular the essential needs of the world's poor, to which overriding priority should be given

2. the idea of limitations imposed by the state of technology and social organization on the environment's ability to meet present and future needs

It might be worthwhile pointing out already at this stage that the original definition is very broad and does not limit itself only to development in the physical or technical sense (as we shall do in the following). Nor does it focus only on environmental considerations, as is often incorrectly implied, but, rather, looks at the whole transformation needed in the economy and society to achieve the distant objectives put forth in the report. Rather than trying to predict our future, the World Commission report is trying to analyze the possibilities for, and obstacles to, a path of sustainable development and long-term environmental strategies.

The conclusion to be drawn from the report is, in my opinion, not one of relief nor of immediate doom or despair. To me, it simply indicates that what we badly need is a serious discussion about the possibilities and technological limits for sustainable development. It is no longer sufficient to treat the problems on a limited local level; the couplings to the environment must be considered more carefully. And by environment, I refer not only to living nature, but to society as well. Only through such a holistic approach can we hope to forsee the long-term effects of different technological choices and consequently direct the development in a desired direction.

An efficient strategy for development, of course, also presupposes that we know where we want to go, which we currently do not. This lack of a coherent development strategy and the fact that, in many cases, we even lack a goal is, in my opinion, the real problem of our time. In many parts of the world, we seem to have reached a level of material wealth which satisfies and, indeed, sometimes even surpasses our needs. Other parts of the world face enormous problems of poverty, even starvation, which seem almost unsolvable within the present mode of continuing growth. It would be wrong to accuse, as is frequently done, technological progress for the uneven distribution of wealth, or even for the environmental problems, as environmental stress often is the result of the growing demand for scarce resources. Environmental problems can, and do occur in regions of low industrial activity as well as in the industrialized part of the world.

From a technical point of view, we need to redefine the strategy by which resource utilization and economic activities can be adjusted to the carrying capacity of the environment. This search for a common strategy would certainly be less difficult if all the problems of economy and environmental stress could be solved in a way that would leave all parties better off, but this is not the case.

One should not, as is often done, let oneself be carried away by idealized parallels to biological life; despite some attractive conceptual similarities, there are important differences as well — differences at a very fundamental level which we will discuss in the following chapters. However, all of the differences are not to our disadvantage; one of the positive ones is that we humans have the ability to at least imagine possible future scenarios. This ability, together with the rapidly increasing capabilities for computer simulation of complex systems provides us with very powerful tools for decision making. Of course, such tools are not creative in themselves, but they are useful in evaluating the different possibilities. Organisms adapt to the environment, they do not foresee changes, and this sometimes leads to disastrous consequences for subpopulations. Humans can perhaps avoid such a passive attitude or at least actively plan ahead and try to avoid it.

The situation is indeed very complex, and looking at the overall picture, it is not clear at all that the present technological priorities are the only acceptable ones, or indeed the right ones, for that matter. Although the rapid development of technology has brought in its wake a number of unanswered questions as well as new challenges, one should not forget its merits. It is, in fact, only due to the enormous material wealth created by rapid industrial development that we can even afford to discuss some of the questions that occupy our minds today.

We cannot, however, deny the fact that an obvious conflict emerges from the present situation. On the one hand, technological development, as represented by the industrialized countries, seems to propel us toward an ecological disaster. In addition to the acknowledged problems accumulated in the past, we have to add the justified suspicion of developing, but as yet unknown threats. On the other hand, it seems evident that economic growth is still necessary in order to ensure a decent living for the entire population of the earth. How are we to resolve this dilemma? Or, indeed, can it be resolved? This problem has been examined from many angles. According to von Wright,[10] one can look upon the situation from the point of view of the past development of man as a thinking creature and come to the pessimistic conclusion that there is very little hope for mankind. Technology will run its course and almost autonomously push mankind to disaster. Pandora's box has been opened and nothing will force the evil (spirits) back into the box again.

One can take the biologist's view, as, e.g., the German biologist Neuweiler[11] has done, and try to force man back to his origin as a biological species among other mammals, remove him from his self-appointed position as "an instrument of divine power".

Chance and necessity have transformed man into such an emancipated being. Through his powers of understanding he no longer suffers the consequences of evolution, he directs its course on his own. Our species no longer adapts to the environment, but changes it according to our wishes. Modern technology is a product of nature. Thus, it would be as useless to forbid its continuing use by man as it would be to forbid a bird to fly. However, indiscriminate use of technology is another matter. The conquest and transformation of the earth by a single, knowledgeable species, namely man, is an inevitable consequence of evolution. Man in the 20th century has brought nature completely under his control and can direct the evolutionary process to achieve his own ends. He is truly the crowning achievment of creation, its subject and object, the very incarnation of evolution.

This is our identity in the history of life. We are neither a supernatural being nor a premature birth or mistake of evolution. We are evolution's most independent creation and its logical culmination - so long as we do not destroy ourselves.

In the next chapter, we will add yet another point of view to the discussion, that of a physicochemist. We will look upon the earth and human society as an abstract physicochemical "system" and try to analyze it in light of the natural laws which have proven so powerful in describing natural phenomena in general. One might object that such an abstraction necessarily involves a simplification *ins absurdum* of a phenomenon we already have admitted to be extremely complex, and consequently it cannot result in anything of value. At best, it will result in self-evident truisms; at worst, in misleading arguments in a pseudoscientific disguise.

We maintain, however, that such an analysis, although admittedly oversimplified, has its merits. Through this type of discussion, it is possible to underscore the fundamental parameters governing the relationship between our man-made society and nature, as well as determine at least the theoretical limits for growth, without indulging in the trap of wishful thinking, which is so near when we are discussing a concept as elusive as the future of mankind.

REFERENCES

1. Pough, F. H. The advantages of ecothermy for tetrapods, *Am. Midl. Nat.*, 115, 92, 1980.
2. Hall, C. A. S., Cleveland, C. J., and Kaufmann, R. *Energy and Resource Quality* John Wiley & Sons, 1985, 19 (and references therein).
3. Meadows D. and D. *The Limits to Growth. A Report for the Club of Rome's Project on the Predicament of Mankind,* Universe Books, New York, 1972.
4. de Larderel, J. Optimal Benefication of Global Resources, *VTT Symposium 102*, 1, 19, 1989.
5. Glasby, G. P. Entropy, Pollution and Environmental Degradation, *Ambio* , 17(5), 330, 1988.
6. *OECD Environmental Data, Compendium 1989,* Les Editions de L'OECD, Paris, 1989, 19.

7. Deutscher Bundestag, *Schutz der Erd atmosphäre*, Bonn, 1988, 485.
8. Economic Commission for Europe, Senior Advisers to ECE Governments on Energy, *Projected Energy Developments in the ECE Region Till 2010 and Pollution*, ENERGY/AC.10/R.2/Add 8, Brussels, December 4, 1989 (and references therein).
9. The World Commission on Environment and Development, *Our Common Future, Oxford University Press*, New York, 1987.
10. von Wright, G. H. *Vetenskapen och Förnuftet*, Söderströms & Co., Förlags Ab, Helsingfors, 1986.
11. Neuweiler, G. *Evolution und Verantwortung*, Alexander von Humbolt Stiftung Mitteilungen, Hamburg, 1986, 1.

CHAPTER **3**

Thermodynamics

3.1.   DEFINITIONS

Before we proceed, it may be useful to review a few central definitions and fundamental relations of thermodynamics. No prior knowledge of this fascinating and powerful branch of natural science will be assumed.

It is customary, and we shall later provide ample evidence for the necessity, to clearly define the extent, the borders, that limit and define the "world" we want to study. The part of the physical world that has been defined by our choice of border is called the *system,* everything else in the universe, as far as we are concerned, can be defined as *surroundings.*

A system, that does not exchange matter, but which may exchange energy with its surroundings, is said to be a *closed system.* A system which exchanges matter with its surroundings is said to be *open,* and a system which does not exchange matter or energy is defined as *isolated.*

From the above, it is evident that the earth can, at least to a very good degree of approximation, be separated from the rest of the universe and viewed as a closed system which does not exchange matter with its surroundings, but which, as we shall see, does exchange energy. Strictly speaking, this description is not true, as the earth receives a constant flux of cosmic particles and loses gas molecules through diffusion to the space.

Energy is defined in mechanics as the ability to do work, a definition not totally adequate, as we shall later see when discussing the central importance of energy in our society. Energy exists in two forms: *kinetic energy,* which is possessed by a physical object by virtue of its motion, and *potential energy,* which is possessed by a physical object which feels a *force* (is positioned in a force field).

The force can be mechanical (e.g., a stretched or compressed spring), electrostatic, gravitational (the attraction of the earth on another stellar body), even chemical, in which case we speak of chemical potential energy, which derives its origins from the tendency of atoms in a particular configuration to arrange themselves into a configuration of lower energy, a more stable configuration as we say.

We are, of course, familiar with the notion of work, at least in principle. And we perceive it, quite correctly, as something which is energy consuming. In its most abstract definition, it is defined as the product of the distance an object is moved and of the component of force along this direction of movement. It is a quantity which has a unit of energy, consistent with the initial definition of energy mentioned in the beginning of this chapter. It is perhaps worth mentioning at this point, although it will be evident from what follows, that energy, or work for that matter, as such is by no means synonymous with what we consider useful work.

Further, to correctly describe a system, we need to specify a certain number of properties of the system in addition to energy, such as, e.g., temperature, pressure, and the quantity of material involved. Any function which may be expressed by these parameters is called a function of the state of the system.

The first principle of thermodynamics is familiar to most people. It is the principle of conservation of energy; energy cannot be destroyed, it can only change nature. The increase in the energy (ΔE) of a system is equal to the heat (Q) absorbed by the system and the work (W) done on the system:

$$\Delta E = E_2 - E_1 = Q + W$$

where E_1 is the initial state and E_2 the final state, and the increase (or decrease) in energy is dependent only on the initial and final states, not on the path taken to reach the final state. In its most general form, which perhaps emphasizes better the dynamic nature of the energy concept, the definition of energy states that there always exists a function of state, called the energy of the system, such that its change per unit time is equal to some flow called the energy flow from the surroundings, i.e., the change of energy of the system is equal to the amount of energy received or lost to the surroundings.

The second principle of thermodynamics is often quoted (but less often understood). It postulates the existence of a function of state, called *entropy* (from the Greek word for evolution), which has the following properties:

1. The entropy of a system is the sum of the entropies of the parts that make up the system (for energy, such a property is called extensive in the language of thermodynamics).

2. The change of the entropy (ΔS) can be split into two parts, one (ΔS_e) denoting the flow of entropy due to interactions with the surroundings, and another (ΔS_i) denoting the contribution due to changes inside the system.

The important statement of the second principle of thermodynamics is that the change of entropy inside the system is never negative. It is zero when the system undergoes reversible changes only (changes requiring no energy and infinitesimally small departures from an equilibrium position), but is positive if the system is subject to irreversible changes.

For isolated systems, there is, by definition, no flow of entropy from the surroundings, as there is no exchange of energy or material, so $\Delta S = \Delta S_i \geq 0$ is always greater than or equal to zero.

It is useful to define a few additional thermodynamic functions with some descriptive value, such as enthalpy H. The increase in enthalpy ΔH is equal to the heat absorbed by the system at constant pressure for changes that involve only expansion work.

$$\Delta H = \Delta E + \Delta(PV)$$

Another useful function, particularly for chemists, is called the Gibbs free energy, often denoted by G. The change in Gibbs free energy

$$\Delta G = \Delta H - \Delta(TS)$$

is in a way, a measure of the system's ability to do "useful" work, or its tendency for spontaneous changes. When $\Delta G = G_2 - G_1 = 0$, the system is in equilibrium between the two states G_1 and G_2, and no spontaneous change will occur. (It also follows, as for the other functions previously defined, that for a cyclic process returning to its initial state, $\Delta G = 0$.)

Particularly for open systems involving the exchange of matter, it has proven useful to introduce the concept of chemical potential, usually denoted μ_i and defined as:

$$\mu_i = \left(\frac{\partial G}{\partial n_i} \right)_{T,P,n_j} = G_j$$

The chemical potential μ_i of substance i is equal to the rate of change in the Gibbs free energy of the system when changing the number of moles (n_i) of this component when temperature, pressure, and the number of moles of the other components are held constant.

The significance of the chemical potential of a component is that it reflects the increase in capacity for types of work other than the pressure-volume work of the system per mole of substance added (for infinitesimal additions at constant temperature and pressure). In this way, the chemical potential is equal to the partial molar Gibbs free energy, G_i.

Based on these kinds of abstract definitions and considerations, a powerful branch of science called thermodynamics has been built. Consider a reaction to which, through the material flow, we bring an enthalpy content H_i and entropy S_i while the reaction products correspondingly remove H_f and S_f. If the reaction is irreversible, there might also be an entropy "production" corresponding to S_{irr}. An amount of heat Q may also be brought into the reaction from the exterior (Figure 1).[1]

The temperature T_R at which the reaction will occur spontaneously can be determined by finding out the conditions for reversibility, at which temperature the heat (Q) supplied, equivalent to the enthalpy change $\Delta H = H_f - H_i$, matches the entropy change $\Delta S = (S_f - S_{irr} - S_i) = Q/T$.

$$T_R = \frac{H_f - H_i}{S_f - S_i - S_{irr}}$$

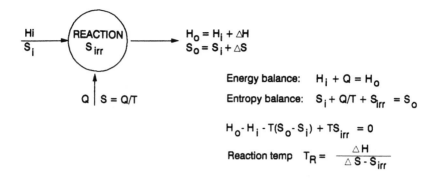

Figure 1. Schematic representation of thermochemical process proceeding in one stage.[1] Temperature must be raised to T_R to match the enthalpy requirement $\Delta H = Q$ with the entropy requirement $Q/T = \Delta S$.

This relation is illustrated in Figure 2[1] for the water-splitting reaction $H_2O \rightarrow H_2 + \frac{1}{2} O_2$. It also underscores one of the obstacles for the spontaneous occurrence of this technically desirable reaction for hydrogen production. Enormously high temperatures (large quantities of energy) are needed.

Thermodynamics provides descriptions and explanations for many complex phenomena in nature in the domains of chemistry, physics, and biology. We will make use of these arguments in the following discussion when illustrating some of the characteristics of man-made society and the biological system of nature.

3.2.   EARTH AS A THERMODYNAMIC SYSTEM

The thermodynamic laws quoted above are general, and as far as we know, there are no exceptions to them. At first sight, it may appear difficult or even contradictory to accept that the laws of entropy maximization, i.e., the tendency of all systems, if left alone, to strive toward a state of maximum disorder, can be compatible with such a complex world as ours. After all, as has been said with doomsday prophecy, "the laws of thermodynamics allow only death".[2] Later on, we will return to this paradox, which has intrigued scientists for centuries and which is still at the forefront of scientific research. It seems that we are here touching upon the very secret of biological life, self-organization from disorder.

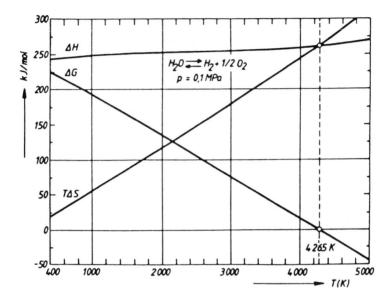

Figure 2.   ΔH, ΔS, and ΔG are reaction enthalpy, reaction entropy, and reaction Gibbs free energy of water splitting, respectively. Note the decrease of ΔG with increasing temperature T.

Let us for the moment return to the simple definitions above and state that the earth can be viewed as a closed system. It receives solar energy, part of which is returned to space, but no matter is exchanged. As mentioned earlier, the latter is not totally true, but for the moment we do not need to worry about the small amount of material (stellar particulates) occasionally entering our atmosphere or gaseous substances diffusing out from it.

Within this closed system, enormous quantities of energy and materials are continuously transformed. Under the influence of solar radiation, biomass is synthesized from simple constituents (water and carbon dioxide) to form complex structures (plants, biological life) which in turn support new forms of more complicated biological life. Everything is used and reused over and over again. Through the carbon cycle, the nitrogen cycle, and, finally, the water cycle, materials are purified and reused in a sustainable manner (Figures 3 and 4). What lessons can we draw from this example of sustainable development to shape our own society? For one thing, the growth of every new population always happens at the expense (decline) of another. Such is the harsh law of

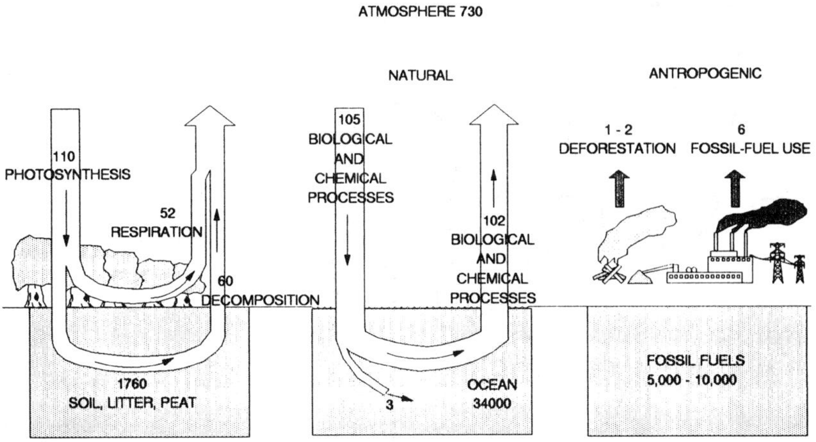

ATMOSPHERE 730

NATURAL ANTROPOGENIC

110
PHOTOSYNTHESIS

105
BIOLOGICAL
AND
CHEMICAL
PROCESSES

1 - 2
DEFORESTATION

6
FOSSIL-FUEL USE

52
RESPIRATION

102
BIOLOGICAL
AND
CHEMICAL
PROCESSES

60
DECOMPOSITION

1760
SOIL, LITTER, PEAT

OCEAN
34000

3

FOSSIL FUELS
5,000 - 10,000

Figure 3. Carbon cycle of the biosphere.[3] After Bolin, B., University of Stockholm.

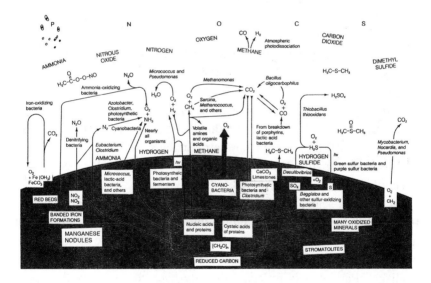

Figure 4. Within the closed system of the biosphere, enormous quantities of materials are continuously transformed for reuse through complex cycles. (From Margulis, L., *Symbiosis and Cell Evolution,* W. H. Freeman and Co., New York, 1981. With permission.)

sustainable development of nature: no democracy, no happiness for everybody, only survival of the fittest. But let us start with a closer look at the energy balance of mother earth (Figure 5).

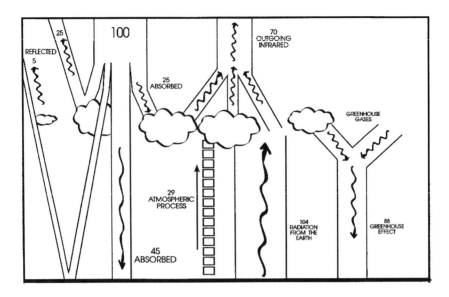

Figure 5.   Energy balance of the earth's incoming radiation energy (178,000 TW).[3]

Solar radiation actually provides the earth with a continuous flow of energy, corresponding at the equator to 1353 W/m^2 (absorption and scattering reduce this value to about 1000 W/m^2, and clouds can reduce it as low as 100 W/m^2). Seasonal and geographical variations are large; however, the June mean value of radiation on a horizontal surface varies between 4.5 kWh/m^2 • d in northern England to 8.5 kWh/m^2 • d in southwestern Spain, the corresponding December values being 0.5 kWh/m^2 • d and 2.5 kWh/m^2 • d.[4]

On a global scale, solar radiation provides the earth with an energy flow of 178,000 TW. To put things into perspective, it may be worth noting that in 1986 the global power requirement of the entire human population amounted to 9 TW; today, it is closer to 10 TW. Where does it all go? Part of it is reflected directly back to space by the upper atmospheric layers and by the earth's surface, and some is lost as longer-wavelength heat radiation from the warmed surface of the earth. Of the total incident flux of energy, only about 0.12% (3.1 • 10^{21} J; 0.3% inland, 0.07% on ocean) is used to support the organic life on earth. Although not very efficient, it nevertheless is about ten times more than the present human annual energy consumption.

The efficiency of photosynthetic activity in plants with respect to incident light varies with species, but the order of magnitude of energy utilization is only in the range of a few percent (Table 1).

The assimilation of energy from food in animals varies from 70 to almost 100% (98% for hummingbirds that feed on digestible sugar).[5] Energy conversion in nature does not appear to be very high in all systems, but we can see a trend toward better energy economy in higher organisms. Clearly, energy economy has been one criterion that natural development has followed, but it has not been the dominant one; in fact, it could not have been. There is one more mystery we must deal with. In view of the simple thermodynamic relationship presented above, one would think that a system should strive toward an equilibrium state of minimum potential energy and maximum entropy, i.e., maximal disorder. This is true for an isolated system, but for a closed system such as the earth that exchanges energy with its surroundings, the situation becomes more complicated.

An equilibrium structure can be seen as a final state which, once attained, can be cut off from its surroundings and conserved, eternally separated, without further interaction with its environment. But the biosphere on earth is not an isolated system, nor is it in equilibrium. Living matter on earth (plants, animal cultures, or bacteria), when cut off from their surroundings (isolated from their environment), strive toward a thermodynamic equilibrium and die. They are open systems that derive their "strength", their very reason for existence, from the exchange of material and energy with the environment. Note that the choice of system boundaries determines whether a system is isolated, closed, or open. The biological cultures on earth are open systems, as they exchange matter and energy with their surroundings; the earth itself is a closed system, as it exchanges only energy with its surroundings.

The consequences of nonequilibrium thermodynamics are far reaching. Only recently, through the pioneering work of the Belgian physicist Ilya Prigogine and his school, have we come to realize that the existence of such highly organized matter as biological life on earth is not in contradiction with thermodynamics, that, in fact, when a system for one reason or another is pushed very far from its equilibrium conditions, such self-organization from chaos is a necessity rather than chance. The

Table 1
Assimilation of Energy Received by Various Organisms[5]

Plants (wild strawberry)	0.85% of solar
Fish (Pacific salmon)	85%
Snake (*Hebrodon platzarhius*)	89%
Human	75%

Source: Hall, C. A. S., Cleveland, C. J., and Kaufmann, R., *Energy and Resource Quality*, John Wiley & Sons, New York, 1986.

maintenance of such a state far from equilibrium, however, requires a constant flux of energy and matter. Rather than striving toward a state of minimum absolute energy and maximum entropy, it turns out that biological systems try, in the most energy-efficient manner possible, to maintain the high order of complexity necessary for their existence. This behavior is not a peculiarity of biology; it may occur in other far-from-equilibrium systems as well, such as turbulent flow or similar complex phenomena. Such a state, as it appears to be stable over time, is called a stationary state, and is characterized by a minimum entropy production per unit time compatible with the boundary conditions of the system. Such a state extracts energy and material from the surroundings, in order to maintain or even increase its internal order, and exports disorder, waste, and heat. In order to be as energy efficient as possible, the system tends toward a state, compatible with the given boundary conditions, in which this waste exportation (entropy production) is at a minimum.

3.3.   THERMODYNAMICS OF THE TECHNOSYSTEM

Man has always used his ingenuity to cope with nature, but we generally regard the industrial period as having begun essentially in the 19th century with the introduction of the steam engine. Why is that so? Surely the discovery of a 7.5-hp steam engine alone cannot have been all that dramatic. And still it was — well, perhaps not the machine itself, but the principle that was introduced with the steam engine. Up to that point, all the technical driving devices used by man had been mechanical. A mechanical machine gives back as work the potential energy stored in it or received from the surroundings. The driving force is mechanical and so is the effect produced. And even more important, at least in principle, such a process (work cycle) is reversible; the machine can be restored to its initial state, after having performed work, by a simple reversal of the process.

With the introduction of the steam engine, a fundamentally new concept emerged, the heat engine. The real discovery was the insight that heat can produce work, or perhaps even more generally, that the transformation of matter can be made to produce useful work. Fuel burns and produces heat, heat produces the vaporization of water to a state of larger volume, and the volume increase is used to produce a forced movement, work. The wheels of industry started turning.

The discovery of the first heat engine by James Watts (1736–1819) generally marks the start of the industrial era. Unfortunately, as we have discovered only recently, it also started the count for the many negative processes that we shall have reason to come back to later. I am, of course, referring to the recent evidence of many cumulative degradation processes that now threaten our natural environment. The origin of these are the same as that for which we praise the industrial system, the liberation of man from the production limits imposed on him by his limited muscular power and his relative vulnerability.

A mechanical machine only transfers motion, while a heat engine transforms heat into work; in a manner of speaking, it actually "produces" work. Of course, the first principle of the conservation of energy has to be obeyed; while the fuel burns, chemical energy is transformed into heat and, eventually, the heat is transformed into work in the heat engine. But there are further limitations. Not all energy-conserving transformations are allowed in a heat engine.

It was mentioned earlier that a mechanical device transforms potential energy into work and that the process is reversible. This statement represents an idealization of the process; in general, some of the potential energy is transformed to heat through friction and this part is irreversibly lost.

The limitation of a heat engine is that it works on the temperature difference. In a way, it transforms one difference into another; when the differences are evened out, the machine ceases to operate. Moreover, the efficiency of a heat engine, i.e., the efficiency by which a quantity of heat can be transformed to work, is dependent on the temperature difference over which heat is made to flow. This dependence was first expressed by a relation developed by the French engineer Carnot

(1796–1832) and later derived theoretically by Clausius (1822–1888). The temperature dependence of the efficiency is remarkable in its simplicity and generality, as it is totally independent of the type of heat engine employed.

The maximum efficiency of a Carnot cycle can be expressed as:

$$\eta = \frac{W_{max}}{q_2} = \frac{T_1 - T_2}{T_2}$$

where W_{max} = the maximum amount of work and q_2 is the heat absorbed (fed to the machine) at the higher temperature. η signifies the efficiency by which heat can be turned to work when T_1 = the temperature of the heat source and T_2 = the temperature of the heat sink.

This relation illustrates the intuitively obvious fact that we cannot extract heat from a colder body and use this heat to do useful work. The calorific content of the oceans, although immense, cannot be used to perform useful work unless a temperature difference is also involved; the mere quantity of calories, no matter how big, is of no use. The Carnot relation puts a theoretical limit on the efficiency of a heat engine; the real efficiency is always lower, sometimes considerably lower, due to inevitable technical heat losses in the process. The final efficiency of a real heat engine is a product of two factors. The first one is dependent on the technology used, and expresses the fraction of heat actually available to the heat engine out of the total quantity delivered to the process. In burning fuel, for example, only a fraction of the chemical energy contained in the fuel is delivered as useful energy to the machine; the other part is lost as waste heat with the flue gases, etc. The second factor in the total efficiency product is determined by the temperature at which the heat engine operates and the temperature of the available cooling system used as a heat sink.

Thus, if W_{max} is the maximum work that can be obtained from a quantity of heat q_2, we have:

$$W_{max} = q_2 \left(\frac{T_1 - T_2}{T_1} \right)$$

where T_1 is often the ambient temperature T_0.

$$W_{max} = q_2 - q_2 \left(\frac{T_2}{T_1} \right)$$

The second part in the expression represents the fraction of the heat which is lost (with respect to work, it can still be used for other useful purposes such as heating).

Much of today's engineering deals with efforts to minimize the loss term in the expression or to find useful means of utilizing the "waste heat", such as in cogeneration. In cogeneration, the waste heat from an electricity power plant is used for district heating.

The major part of our industry today is driven by so-called heat engines of different kinds, and they are all subject to constraints imposed by the relations above.

From these considerations, it must be evident to everybody that energy is a vital quantity for both the ecosystem and the technosystem, and that energy considerations alone do not suffice to describe the characteristics of a system, nor can conservation of energy be the only guideline for biological or technical development. Further, it is perhaps worth emphasizing the fact that although energy cannot be lost in the strict sense of the word, it can be "degraded" (dissipated), and in that sense made unavailable for further use.

It was mentioned earlier that the earth can be viewed as a closed system exchanging only energy with its surroundings, and that the many biological systems existing on earth can be described as open systems exchanging both energy and material with their surroundings.

The technosystem of today can also be pictured as an open system. In a typical industrial process, the raw material or a combination of raw materials are fed into a process in which they are brought to react with each other to produce a desired product following the simplified relation of Figure 1. Unfortunately all reactions generally entail a certain amount of undesired byproducts of the reaction, i.e., waste. Most often, the desired reaction does not proceed spontaneously and we need to add energy to the process in order to shift the equilibrium toward the desired direction. In such a case, part of the energy added is inevitably lost (dissipated), generating waste heat, which instead of being utilized, often poses an environmental threat in itself.

Although the various biological subsystems making up the ecosystem and the technosystem can be viewed thermodynamically as open systems, there are still fundamental differences between them. We mentioned already that a biological system tries to minimize entropy production, i.e., energy dissipation, while maximizing its growth (or perhaps its conditions for survival) subject to the available energy constraints — sunlight, food, or space limitations. An industrial process, again, is concerned with maximizing production given certain economic constraints. But in addition, and often overlooked, the separate biological systems are interwoven with each other and eventually make up one big, closed ecosystem, the biosphere of the earth. The various separate processes making up the technosystem, on the other hand, are not truly dependent on each other. There is, of course, the weak economic interaction, but that does not make them thermodynamically interlocked. Thus, the separate industrial processes remain theoretically and practically open systems, interacting with the ecosystem more or less uninhibited (until now) by any global laws (at least in the thermodynamic sense).

One further fundamental difference between the two systems worth mentioning is the enormous difference in energy intensity. The energy intensity may not be an absolute necessity of the technosystem itself, but, rather, a consequence of human ingenuity. Nevertheless, the fact remains that industrial production today consumes energy and produces goods at a rate which far surpasses any biological system. Table 2 illustrates these enormous differences in energy densities (energy used per surface area unit) between different human activities. An oil refinery, representing a typical, condensed industrial plant, uses 1 million times more energy than a "biological" activity such as farming. Of course, product outputs per unit area from the different processes are different too, as seen from Table 2, but as we know from the laws of thermodynamics, no process is perfect in its energy use, and the amounts of waste produced are unfortunately proportional to energy use. Here lies one of the major difficulties facing our technological society: it is not the inefficiency of our industrial processes compared to natural biological processes, but rather the converse that is one of the major causes of environmental pollution. The rate at which waste is produced is simply much too high. It is true that the ecosystem produces waste too, but not nearly at the same rate, and certainly not at the same rate per area. Moreover, biological "production" units are frequently mo-

Table 2
Order of Magnitude Energy Densities of Some Human Activities[1]

	GJ/ha/year	Product flux[a]
Hunter-gatherer society	0	30–50 kg/ha/year
Sheep farm	0.6	100–150 kg/ha/year
Mixed farm in industrial country	10	35 ton/ha/year
Intensive crop production	15	5–10 ton/ha/year
Intensive fish farm	200	100–300 ton/ha/year
Singapore state	950	—
Manhattan	20,000	—
Oil refinery	1,000,000	1–2 Mton/ha/year

[a] Approximate values for product quantities.

bile, and move away from their waste. When this is not possible, they tend to form a culture of different species, where the waste products from one are utilized by others, eventually leading to something similar to a closed system.

One might ask why we do not do the same, hook one process to the other until all the waste streams are utilized. We may well have to, the present trend toward recycling is a step in this direction, but it is not very easy to create such a system. One must remember that Nature has had time for many mistakes in the trial and error procedure it has performed during the 200 to 300 million years that life has existed on earth. At the rate we are doing things, there is not much room for error.

Such an integration also leads to a complex system in which it is difficult to evaluate the effect of changes. This complexity is the very reason why it is so difficult to foresee the harmful effects of (even small) environmental changes.

If the difference between closed and open systems is only a matter of how we define our system boundaries, why do we not just include the waste production in the system and then the system is closed? Indeed, this could be possible (and is sometimes done under the name "ecological lifestyle"); the problem is that at the production rate of a typical industrial process, the amount of waste produced is enormous. The fact that these products often occur in diluted form further exaggerates the problem.

A process can, of course, at least conceptually, be designed to concentrate its wastes, reuse everything possible, and eventually store, within factory limits (system boundaries), only the absolute minimum that cannot be reused. Apart from economic limitations, this is technically feasible; the economic difficulties lie in the fact that the cost of treatment of a low-grade raw material rises sharply with the decreasing concentration of useful material in the waste. In some cases, this type of "closed" process has, in fact, been realized, the most obvious one perhaps being a nuclear power station. Assuming that the waste products from the reactor are stored on site (which is feasible) and that there are no leaks, such a process could justifiably be considered an example of a closed process. This is hardly the idea of a final solution to environmental problems for those who advocate closed industrial cycles, but a good example in that it illustrates one crucial point in the argument; the more we concentrate the unwanted fractions of our industrial waste products, the more potentially malign these wastes tend to become.

3.4.   THERMODYNAMICS AND ENERGY IN SOCIETY

From what has been said, the central role of energy in the industrial system, or in the biological one for that matter, must be evident. Energy is the ultimate commodity; through the use of energy, we can produce food, construct shelters, organize transport systems, and create order.

It has been estimated that the total human population on earth, today some 5 billion, will level off at around 14 to 16 billion by the middle of the next century. This assumption is based on the observation that the rate of nativity in nations declines as the living standard of its inhabitants rises. The projections differ somewhat due to different assumptions for industrial and economic growth rates in the various parts of the world. Despite this tripling or quadrupling of the total population, there is no indication that food, the typical limiting factor for biological systems, would be an absolute limiting factor for humans. In fact, some studies indicate that the earth's capacity for food production could be increased by as much as 30 times the present production rate![6] However, a word of caution should perhaps be raised here; factors such as land erosion and groundwater pollution, with the subsequent loss of arable land, may well change these optimistic forecasts in the long run. What is certain is that such a production rate would need high energy inputs on a global level. The dependence of farming output per unit area on energy input is demonstrated by Figure 6.

Energy, as mentioned previously, is the ultimate commodity. It is also the only resource that theoretically cannot be reused; it is degraded and

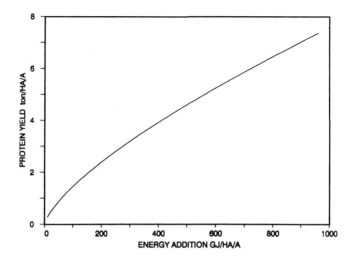

Figure 6. Agricultural protein yield (kg/ha/year) as a function of added energy (GJ/ha/year). (After Slesser, H., Lewis, C. and Edwardson, W., *Food Policy,* 3, 123, 1977.)

lost after having literally been put to work. Thus, it is not surprising that so much attention is devoted to the price and use of energy. In fact, the affluence of a nation is very clearly reflected by its energy use, as expressed in Figure 7, where the per capita gross national product values of different nations are seen to correlate directly with their per capita energy consumption. The lower end of the scale corresponds to muscle power, while the per capita energy consumption of industrialized nations is nearly 100 times more, the equivalent of 100 obedient slaves. It has been claimed that this linear relationship, and in particular the steepness of the slope, need not be a fundamental necessity, that such a dependence is technology dependent (which is certainly true) and that the energy use per capita could be dramatically lowered without affecting the income level, by altering the industrial structure of nations.[7] However, although a shift toward higher energy efficiency (measured as energy/GNP) is clearly visible in the industrialized countries, no dramatic changes have been provided on the national level; the arguments for their existence have been based on extrapolations from isolated cases, and their validity on a more global level remains to be proven.

Let us take a closer look at the energy prospects of our society. The total human population today is about 5 billion. If all these people were

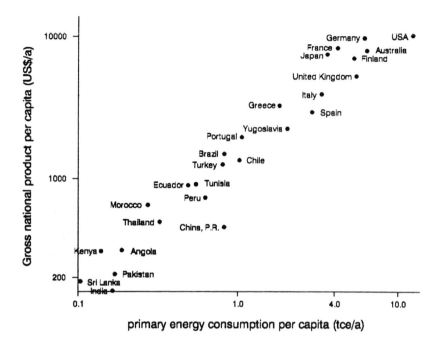

Figure 7. Gross national production per capita as a function of primary energy consumption per capita.

to enjoy the standard of living of the U.S., for example, global energy consumption would increase from the present 7.3 Gtoe/year to 40 Gtoe/year, an increase of nearly sixfold. Such a transition will not happen suddenly, and the population will continue to increase. If we take the estimated leveling off value for the population as a reference, then energy consumption by the year 2050 will have increased 12 to 15 times compared to the current level. Clearly, this is too much to cope with given our present-day technology or resources.

3.5.   THERMODYNAMICS AND ENVIRONMENTAL POLLUTION

Although the flow of energy is of fundamental importance in thermodynamics, and despite the fact that energy is the driving force in all industrial activity, depletion of potential energy resources is not the only threat to continued industrial activity. In fact, we are forced with several important problems. On the one hand, we have (albeit briefly) been shaken by the chilling prospect of a global energy shortage; on the other hand, we are constantly reminded of the need for economic growth to improve the desperate situation of many developing nations. Add to this the ever-accelerating flow of information on severe pollution problems over almost the entire world due to industrial activity, and one realizes that it takes a real evolution optimist to believe that there is a way to possibly change things for the better. Fortunately, there still are a few of us left who believe that positive action can be taken, but it is clear that such changes require more than the benevolent thoughts of a few optimists. Drastic measures are called for, and the solutions are not evident. As in the case of energy conservation, where a return to manual labor would increase total energy consumption, unless a serious deterioration of living standards is accepted, there is no way back, as far as environmental pollution is concerned, unless we accept a return to materially much simpler, and for many, much more inhumane conditions of living.

Pollution prevention requires investment and energy, and thus puts added pressure on the requirements for economic growth. Still, we have no choice! In retrospect, one cannot avoid realizing that we have, in many cases, been slow to act; the writing on the wall should have been apparent long ago. We are, through the indefatigable efforts of entropy, spreading our effluents over the entire earth. Take, for example, the Rhone and Danube rivers. Once the freshwater streams of large parts of the most densely populated parts of Europe, these rivers are today hardly more than gigantic sewers (Figure 8). Who's to blame? Industry? Certainly, industry is responsible for the more flashy headlines as far as bad publicity goes, but in quantities of effluent, the biggest polluter by far is agriculture and the careless use of fertilizers (Figure 9), not to mention "domestic" effluents. Again, it is the scale of human activity, compared to the receiving capacity of the surrounding environment, that is the origin of the problem — not the quality of the effluents as much as the quantity. A particularly illustrative example is provided by the carbon dioxide problem. In itself, carbon dioxide is as harmless as any chemical can be; it is a major and necessary constituent in the biosynthesis of living matter, yet its increased content in the atmosphere, from a mere 200 to 350 ppm over a period of a few hundred years, is considered to be the most serious environmental problem of the near future.

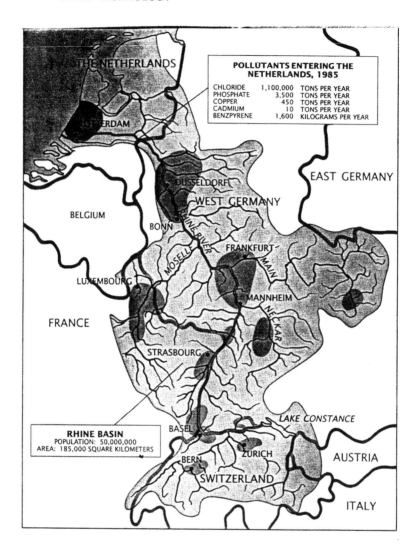

Figure 8. Rhine River drains a vast basin in four countries — Switzerland, West Germany, France, and the Netherlands — as it runs 1,320 km from the Alps to the North Sea. The basin is heavily industrialized, and the river accumulates and transports into the Netherlands a heavy load of pollutants; since 1980, the amounts of some pollutants have been reduced. Now, the four countries are cooperating in a Rhine Action Plan intended to improve the quality of the river's water. The primary effort is to institute recycling within industry as a substitute for after-the-fact, "end of pipe" treatment. (Source: Rivière la, M. *Scientific American,* September 1989(53). With permission.)

The halogenated hydrocarbons, often incorrectly called freons (actually freon is the popular name for only a small subclass of these compounds), could be cited as another typical example of a class of

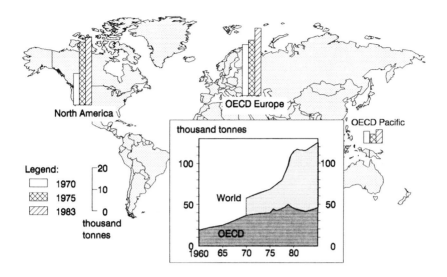

Figure 9. Apparent consumption of commercial fertilizers (nitrogenous, phosphate, and potash fertilizers). (Source: OECD Compendium, 1985.)

chemicals that have become a problem, not because of their reactivity, but, oddly enough, because of their lack of reactivity. Due to their chemical inertness, these compounds resist chemical degradation and migrate to the stratosphere, where they participate in complex reactions induced by the energy-intensive ultraviolet radiation, resulting in a degradation of the thin layer of ozone that normally protects us from excessive intensities of ultraviolet radiation. Ozone is vital as a protective shielding at high altitudes in the stratosphere, but when present closer to the earth's surface, it is believed to be one of the chemicals responsible for the much talked about forest decline in middle Europe, particularly in the mountainous regions.

These are only a few examples of the complexity of the interaction between man and nature; actually, it is a question of the right products in the right place, and in the right quantities.

Unless particular precautions are taken, which require energy, the second law of thermodynamics favors an even distribution of all substances (total disorder), and that is what we see happening around us. The difference in entropy in keeping a number (C) pure components entirely separated or totally mixed together is:

$$\Delta S = -R \sum_{i=1}^{C} N_i \ln N_i$$

where N_i denotes the mole fraction of component i in the final mixture. As N_i ranges from nearly zero to nearly unity, the logarithm is always negative and the entropy of mixing is always positive, as it should be, indicating an increase in entropy for the disordered state. The inverse operation of separating mixed components requires energy and ingenuity. Effluents spread out and are diluted into the air, water, and earth. When the quantities are small, dilution is a rather benign and convenient way of getting rid of wastes, provided that the receiving environment can accept and eventually degrade the wastes without suffering damage. This dilution principle is the disposal method most often used by nature itself. However, when the quantities of waste grow too large, either locally or on an absolute scale, for the environment to cope with, dilution disposal results in disaster. The reversal of the mixing process does not occur spontaneously, but requires energy and ingenuity to effectuate. Theoretically, the energy (entropy) of demixing is relatively small, but in practice, our methods for separation are rather inefficient, resulting in the investment of considerable amounts of energy and effort in the separation processes. In fact, the major part of the chemical industry is concerned with the separation of various compounds from each other. In practice, separation processes need substantial amounts of energy, and the energy requirements increase rapidly with the required degree of purity. Thus, although energy is required to keep effluent streams from mixing into the environment, this energy requirement is much smaller than the amounts needed to undo the mixing and separate the constituents again.

An illustrative example is the problem of desalination of water, an important concern in many arid regions. Although the theoretical energy requirement for demixing 1000 l of seawater is only about 4 MJ, we have needed some 280 MJ, depending on the process used, to produce desalinated water. (See Chapter 5, Table 2, for progress in this area due to modern technology.) The situation is similar in many other separation processes. Why is this so? Part of the answer is insufficient technology, and that is the driving force for further technical development as well as the source of inspiration to a certain optimism that things can still be improved. But this does not necessarily mean that, in the real world, we can ever hope to come close to the theoretical figure which is valid for an ideal case of noninteracting components.

In reality, all our separation processes depend on relative differences between the species to be separated, as opposed to the idealized situation which assumes an absolute way of identifying the different components. As a consequence, all technical separation processes result only in a difference in the distribution of the component to be separated between two phases, rather than a perfect separation based on discrete on-off choices between each component particle. The two phases can be liquid-vapor, as in distillation, or liquid-liquid, as in extraction, leaching, or washing. In order to increase the concentration difference between the two phases in the separation process, we need to either change the relative ratio of the two phases brought into contact, i.e., dilute one of the phases, or repeat the process over and over again, which is more efficient. In both cases, the volume of the stream into which the substance has been extracted increases, resulting in increased recovery costs. The great energy demand related to separation processes is due to the large volumes of solvents that have to be evaporated and recovered. Alternatively, if we speak about a washing process, the large quantities of dilute effluent streams constitute a serious waste problem if left untreated.

An inevitable consequence of the above considerations is that if we are forced to utilize poor raw material resources, e.g., mineral resources that have a low concentration of the desired compound, this will result in rapidly escalating recovery costs (Figure 10). In fact, at the limit of very low concentration, the energy needed to extract the final traces of a substance, desired or undesired, will be infinite. The practical limit of technoeconomic feasibility will, of course, be reached much earlier, putting a limit on the philosophy of "more from less" sometimes advocated in discussions concerning the limits of growth.

3.6.   TOWARD A THERMODYNAMICALLY SUSTAINABLE DEVELOPMENT

Thermodynamically, the earth is an open system exchanging only energy with its environment. Even if we limit our system to encompass only the biosphere of the earth, this holds nearly true. The biosphere itself, however, is composed of a multitude of open subsystems

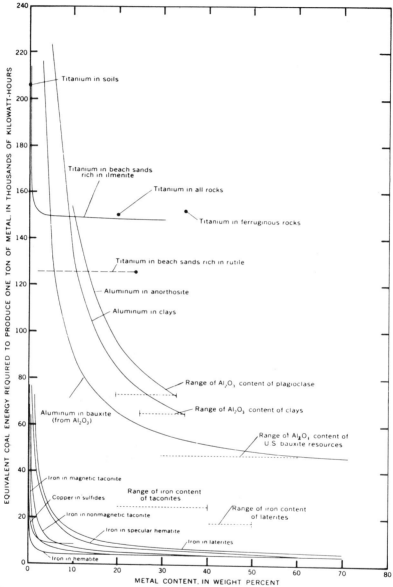

Figure 10. Energy required to produce metals from ores of varying metal content. (Source: Hall, C. A. S., Cleveland, C. J., and Kaufmann, R., Energy and Resource Quality: *The Ecology of the Economic Process,* John Wiley & Sons, New York, 1986. With permission.)

exchanging energy and matter with their surroundings. The man-made technosystem or industrial establishment forms one such open sub-system, extracting raw materials from its surroundings and emitting waste products (Figure 11). The interaction of the technosystem with its

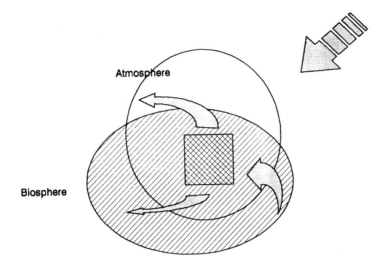

Figure 11. Present technical society in conflict with Nature.

surroundings is assuming such proportions today that it threatens the very existence of the biosystem hosting it. We are literally cutting off the branch upon which we sit. The question raised more and more loudly is, what can we do to alter this journey toward disaster? Sustainable development is the new catchword in this context, but what does it really signify? According to the language of thermodynamics that we have developed here, there are two options. Either we adapt the technosystem to suit the environment (Figure 12), in a way following the same strategy nature has used, or we close off the technosystem from the biosphere (Figure 13). No other options exist. Both choices have their drawbacks. The first one severely limits the technological options available, and perhaps more important, it puts very strict limits on the volume and local concentration of industrial activity, which already have been surpassed in many cases. Adoption of this strategy for technical development would involve significant structural changes of both a technological and a sociological nature. The second option of a closed technosystem, apart from the fact that it would definitely alienate man from his natural cradle, would involve serious technological challenges, and most certainly a significant increase in energy consumption due to the many additional separation requirements involved, when all the waste streams have to be processed further.

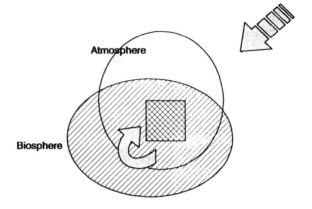

Figure 12. Closed technological system.

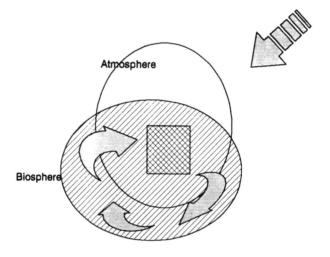

Figure 13. Open system harmonized with the environment.

Biotechnology has made impressive progress over the last years, but it remains to be demonstrated that it can actually sustain industrial production at the level necessary to meet all the expectations put on the industrial establishment of today. The doubts are due not only to the insufficiency of the present level of technology, but lie deeper, in the very nature of biotechnological processes, the driving mechanism which as mentioned earlier, is maintaining order through the minimization of entropy production, not maximal efficiency in production rate. Further, it is worth mentioning that the mere fact that a process is biological at its origin does not necessarily make it environmentally benign; in fact, we have many examples of harmful biological products.

It is often claimed that new technology can change the present vicious cycle of increasing energy consumption with the increasing level of welfare, and this may be true to a certain extent, but it is not clear that an uncritical optimism in this respect is well founded.

Knowledge and information can compensate for energy-consuming operations. Intelligent information systems can save a lot of energy and raw materials in processes or in society as a whole, but these sophisticated systems rely on a complicated infrastructure, the construction and maintenance of which requires investment and energy. Today, a message can be rapidly transmitted over nearly the entire world by means of a sophisticated telecommunications network of computers and satellites coupled to local infrastructure, but if the cost of establishing such a network is evaluated, it is not at all clear that in economic terms it will compete favorably with the ancient messenger custom if the purpose were to send only a few messages. The real gain is the rate of information per unit time that is made possible by new technology (although one might seriously wonder at times whether the quality of all information today merits transmission). The same may be true for many operations, in terms of straight labor, human labor might still be very competitive in its versatility were it not for its low output per unit time.

The average human power output is 100 W, barely sufficient to keep a modest electric lightbulb glowing — not much to brag about in our industrialized world, where every human is used to spending about 100 times more for his everyday life.

REFERENCES

1. Sizmann, R. Solar driven chemistry, *Chimia,* 43, 7, 1989.
2. Prigogine, I. and Steugers, I. *Order out of Chaos: Man's New Dialogue with Nature,* Bantam Books, 1984.
3. Bert Bolin, University of Stockholm (Figure 3); Lynn Margulis, *Symbiosis and Cell Evolution,* W. H. Freeman and Co., New York, 1981. (Figure 4).

4. Grathwohl, M. *World Energy Supply: Resources, Technologies, Perspectives,* Walter de Gruyter, Berlin, 1982.

5. Hall, C. A. S., Cleveland, C. J., and Kaufmann, R. *Energy and Resource Quality,* John Wiley & Sons, New York, 1986.

6. Hall, D. O. Food versus fuel, a world problem? in Proc. 10 CD/UNESCO Conf. Photochemical Conversions, Ecole Polytechnique Federal Lausanne, June 1983.

7. World Commission on Environment and Development, *Our Common Future,* Oxford University Press, New York, 1987.

CHAPTER 4

Energy

4.1. THE GLOBAL ENERGY SITUATION

From thermodynamic considerations, it is not surprising that the supply of energy has such a central role in the evolution of human society. As a rule up to now, the energy consumption of a nation directly reflects its level of industrial activity. The connection between energy use and the economy, and hence wealth, of a nation is less evident but nevertheless real. The details of the connection between a nation's wealth measured, e.g., by its GNP or GDP and its total energy consumption are unclear,[1] as the individual fluctuations between different industrialized nations are large, but the general trend is clear: This is also borne out by the sad fact that the industrialized world uses 75% of the total primary energy production of the world, although it only accounts for 25% of the total world population. In addition, an increase in the living standard results in an increase in electricity consumption (Figure 1).

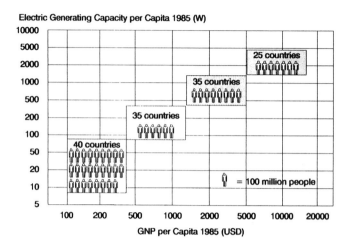

Figure 1. Prosperity and electric energy. (Source: Gluckman, M. J., *Toward 21st Century Coal Processing,* Electric Power Research Institute, 1991.)

The disproportion between energy consumption and population density has also been a source of general concern lately, as we have reason to believe that already the present level of industrial activity exceeds the carrying capacity of the planet, in terms of both the supply of raw materials and acceptance of the environmental load that such an artificial activity puts on the natural surroundings.

Without new technical approaches to industrialization, it is difficult to see how the legitimate aspirations of the less fortunate, but rapidly increasing majority of the world's population can ever be fulfilled. Clearly, this problem is not only technological; 10 years ago, the net money flow from the Northern Hemisphere to the developing nations in the South was $40 billion. Today, it has reversed and the South is transmitting at least $20 billion per year to the North,[3] mainly due to excessive foreign debts, but also in this case the link to technology is there, as much of the borrowed foreign currency has been used to pay for imported oil. It is thus not surprising that the recommendations of the World Commission on Environment and Development report also stress the importance of a satisfactory energy supply for the world[4] in stating the objectives for future development:

1. Sufficient growth of energy supplies to meet human needs (which means accommodating a minimum of 3% per capita income growth in developing countries)

2. Energy efficiency and conservation measures, such that waste of primary resources is minimized

3. Public health, recognizing the problems of risk to safety inherent in energy sources

4. Protection of the biosphere and prevention of more localized forms of pollution

In the report, it is also admitted that the period ahead must be regarded as transitional from an era in which energy has been used in an unsustainable manner. The World Commission on Environment and Development recognizes that an acceptable pathway to a safe and sustainable energy future has not yet been found, but believes that these dilemmas have not yet been addressed by the international community with a sufficient sense of urgency and in a global perspective.

The total consumption of energy today is indeed very high. Every hour, about 500,000 tons of coal are mined worldwide, 2 million barrels or about 300,000 tons of crude oil are extracted, and nearly 200 million m^3 of natural gas are pumped from wells or separated from crude oil. Simultaneously, 1 billion kW of electricity are transmitted from power plants — energy which could light up 5 billion 100-W lightbulbs during that hour (Table 1).

On a per capita basis, these energy flows add up to an annual energy consumption of about 70 GJ, equivalent to about 2.5 tons of coal or 1.6 tons of crude oil for every person on earth.

The figures quoted above give an idea of the magnitude of the total energy consumption of the world, but apart from that, they are not very useful as the energy is not evenly distributed between nations, as mentioned earlier, nor is it evenly distributed on a per capita basis within nations. The actual distribution of energy between different sectors of society within the OECD countries is shown in Table 2.

The prediction of future energy consumption is, despite all efforts, very difficult due to the intricate relationships between industrial structure, economics, and energy consumption.

Table 1
Global Energy Flows in 1987

Energy sources	Energy flows, worldwide		
	0.5 h	1 h	1 year
Coal (t)	250,000	500,000	4.4×10^9
Crude oil (t)	150,000	300,000	2.63×10^9
Natural gas (1000 m³)	100,000	200,000	1.85×10^9
Electricity (kWh)	0.5×10^9	1.0×10^9	9.8×10^{12}

From Okorokov, V. R., *Technology and the Environment: Facing the Future*, Finnish Academies of Technology, 1989:1. With permission.

Table 2
Final Consumption of Energy by Sector (OECD, 1987)

Sector	% of total
Industry	33.26
Transportation	30.24
Other	32.76
Nonenergy use of petroleum products	3.74

After OECD Environmental Data 1989, Organization for Economic Cooperation and Development, Paris, 1989.

The World Commission on Environment and Development works with two scenarios. According to the high-energy scenario, energy production in the year 2030 would involve 1.6 times more oil, 4.4 times more natural gas, and about 5 times more coal than in 1980.

The alternative low-energy scenario, which assumes efficient saving of energy, predicts that the energy demand of the industrialized countries by the year 2020 will have decreased from the 1980 value of 7.3 TWa to 3.9 TWa (about half), while the energy demand in the developing world in the same period will have increased from 3.3 to 7.0 TWa, i.e., more than doubled.

According to the World Commission on Environment and Development, the economic and environmental consequences of the high-energy scenario would be severe, and it strongly recommended forceful actions to increase the efficiency of energy use and the utilization of renewable energy resources.

Improvement of energy efficiency affects both the production and end use of energy, although the potential for further gains is far greater on the demand side than on the supply side. It seems to be more useful to invest in end-use efficiency improvements than on the supply side. Numerous studies confirm this conclusion. For example, a Canadian study of 1984 shows that the real costs to save a certain amount of energy are half of the cost to produce the same amount of energy; similarly, an inquiry of U.S. electricity distributors has shown that in the next 10 years, an additional electricity supply of 30,000 MW could represent an investment of $19 billion, and the cost of avoiding this additional growth through more efficient appliances would be on the order of $6 billion.[5] Shifting the emphasis in energy planning from expanding the supply to improving the efficiency of end use is therefore a central element for consideration. It must be emphasized that the aim of the European Community strategy to improve energy efficiency is to reduce the wasteful use of energy, without necessarily entailing any cut in comfort or production.

The specific energy demand in many industrialized countries has actually decreased substantially during the last 15 years (after the shock of the first oil crisis) through an annual improvement of energy efficiency of 1.7% during the period 1973 to 1983, a reasonably good achievement considering the fact that the annual increase in energy demand (measured in terms of increased industrial production) in the industrialized world over the same period was about 3% (Figure 2). Unfortunately, the net effect of the decrease in specific energy demand and the growth in industrial output has still resulted in an increase in total energy demand (Figure 3).

One must also realize that much of the improvement up to now has been realized by quite simple measures, reflecting poor original energy economy rather than real technical progress. Maintaining the favorable trend in improvement of energy efficiency in the future does, however, call for significant technical development.

Most industrialized countries today have rather ambitious energy research programs aimed at energy saving and the development of alternative energy production methods based on renewable energy resources. However, the inertia in the complex energy system is such that the present fossil-based energy culture will be with us for at least another 50 years before significant new energy production technology can grow to proportions of importance. Of course, this period need not be one of stagnation, intellectually or in economic terms, but rather, a transition period marked by technical development during which

toe/GDP (Million US$)

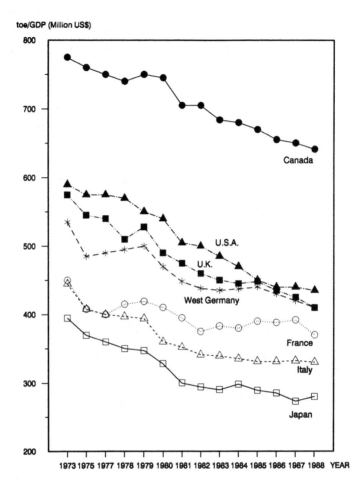

Figure 2. Changes in energy consumption in toe per GDP (millions of U.S. dollars) in some OECD countries.

modifications and improvements of present-day technology will emerge while we wait for scientific and technical breakthroughs allowing the development of a sustainable energy policy.

The first steps toward a clean energy production technology will most certainly continue along the lines of improved specific energy efficiency, i.e., the saving of energy and use of low-value waste heat wherever economically possible. Such improvements have the advantage that they do not suffer from the time lag linked to changing the basic energy economy, but can be put to use almost immediately.

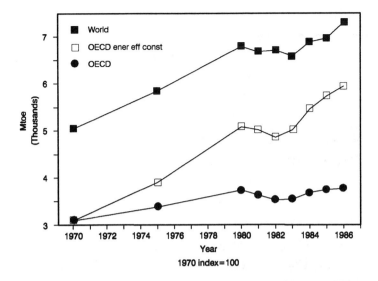

Figure 3. Development of total energy requirements. (Source: OECD Environmental Data Compendium 1989.)

Certainly, fossil-based energy production methods will also have to be developed in a more environmentally benign direction. This, too, can happen rather rapidly, as the basic technology is known and appears economically attractive, but as mentioned earlier, it is only a temporary solution during a transition period toward a sustainable system, as all fossil-based production relies on a finite and nonrenewable material base that can be used only once.

From an environmental point of view, the use of natural gas will increase, as the emissions from burning gas are lower per energy unit than for other fossil carbon resources (Table 3). Also, various gasification processes show great potential for improving the environmental acceptability of many low-grade fuels for producing high-grade electricity.

Natural gas technology is in many respects similar to petroleum technology. One of the advantages of gas production is that it is usually possible to extract 75% of the gas in place without additional measures, and that some of the remaining 25% can be obtained using additional measures. Recovery of the natural gas can also be considered an environmentally necessary action, as releasing methane into the

Table 3
Emissions from Combustion

Fuel	SO$_2$	NO$_x$	C$_m$H$_m$	CO	Dust
Heating oil	23	7	0.2	0.1	1
Gas	—	5	—	—	—
Hard coal	26	7	0.1	0.5	3.5
Brown coal	23	8.5	0.1	0.1	4.5

Kg pollutant/toe fuel

Source: Grathwohl, M., *World Energy Supply,* Walter de Gruyter, Berlin, 1982, 262.

atmosphere increases the global greenhouse effect with respect to the corresponding CO$_2$ release when burning the gas. The applications of natural gas as a primary energy source are such that high efficiencies are generally obtained. It is used primarily for heating in the industrial, commercial, and household sectors.

The transition to natural gas is also an expected development along the path toward a higher H/C ratio of the world energy production which has been characteristic throughout the industrial evolution (Figure 4). In fact, it has been suggested[6] that natural gas will soon be the major primary energy source and remain so for the next 50 years or more, eventually reaching as high as 70% of the world's total energy supply. Further off in the future, we have the truly renewable energy production methods, some of which already are technically feasible and, in fact, actually in use.

The ultimate challenge in renewable energy production is to learn to utilize and technically harness phenomena where the entropy can be diminished locally. In doing so, it will be possible to design effective energy production devices that do not harm the environment.[7]

A simple example of a phenomenon where entropy diminishes is when a lens or reflector is used as a condenser to concentrate light rays on an object or area. Solar energy-based systems aim at tapping the enormous energy flow falling on earth from the sun (Figure 5). The problem does not lie in the lack of basic technology (as will be seen in Section 4.2), as the efficiency of diverse solar energy devices is already quite high, but in the dispersity of the solar radiation. Although solar

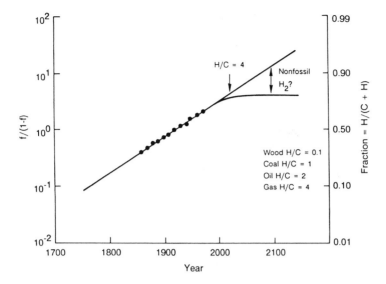

Figure 4. Evolution of the ratio of hydrogen (H) to carbon (C) in the world fuel mix. The figure for wood refers to dry wood suitable for energy production. If the progression is to continue beyond methane, production of large amounts of hydrogen fuel without fossil energy is required (From Ausubel, H. J., *Technology and the Environment,* National Academy of Sciences. Published by National Academy Press, Washington, D.C., 1989. With permission.)

energy is abundant, it is rather evenly distributed over the entire surface of the earth, which calls for cheap solutions for its capture.

Finally, a truly technical solution to man's energy problems is presented by the possibility of man-made fusion technology. The possibility of mimicing the sun's energy production process has long thrilled the imagination of scientists, and through persistent research efforts, this possibility is slowly approaching reality as experimental conditions approach those required for self-sustained fusion processes. The application of large-scale energy production is, however, far in the future, and certainly not without problems.

4.2. THE ENERGY SYSTEM

An efficient energy economy is much more than mere cheap energy production. To understand this, it is important to realize the essence of the energy demand and how this demand is, in fact, an integral part of

INCIDENT TOTAL 178.000 TW

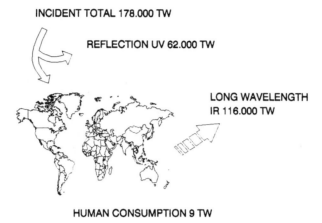

REFLECTION UV 62.000 TW

LONG WAVELENGTH
IR 116.000 TW

HUMAN CONSUMPTION 9 TW

Figure 5.   Solar energy fluxes to and from the earth.

the "energy system", as opposed to the traditional notion of the much more limited "energy sector", as described by Rogner and Britton.[8]

When the phrase "energy system" is used, one is generally referring to the downstream extension of the energy sector to the level of consumer products and services. In this context, the energy system may be viewed as a chain composed of five main segments:

- Energy sources
- Energy conversion technologies
- Energy currencies (fuels)
- End-use technologies
- Services

The components of this source-to-service chain are illustrated by Figure 6.

It is important when thinking of long-term improvements of energy use to consider the whole energy sector, and not just isolated parts of it. The central question to bear in mind is, which is the final service that one intends to provide?

4.3.   FOSSIL ENERGY

As mentioned earlier, the energy production establishment represents such an important part of present-day society that it cannot be modified,

**Example Of An Energy Chain:
From Resource Extraction To Energy Services**

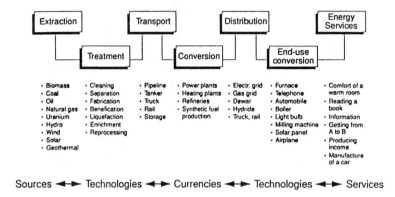

Sources ◄─► Technologies ◄─► Currencies ◄─► Technologies ◄─► Services

Figure 6. The elements of an energy system (From Rogner, H.-H. and Britton, F. E. K., Eco Plan International Report, Paris, 1991. With permission.)

let alone replaced, overnight despite the very obvious requirements placed upon it by diminishing primary raw material resources and increasing environmental concerns.

It has been estimated that the fastest rate at which new technology can penetrate the energy sector is about a 2% annual increase, a figure which is mainly determined by the physical limits of construction and rather independent of economic competition. At that rate, it would take several decades before any new technology could play a major role in the energy production of the world.

It is clear that we have to accept the fact that much of the present energy production technology will stay with us for quite some time, and that the present dependence on fossil fuels will last for at least another 50 years. This means that while working toward technical solutions which in the long run will satisfy the criteria for sustainability, the most immediate actions in relation to clean energy production technology must concern themselves with limiting, as far as possible, the negative effects of the present energy production. In fact, this work has already started, as demonstrated by the substantial reductions of SO_2 and NO_x emissions from power plants, but there are still many things that can, and must, be done to reduce the environmental effects of today's fossil-fueled power plants.

The anthropogenic SO_2 emissions in the world amount to 60 to 80 million tons/year, corresponding to about half of the total sulfur emissions (the main natural sources being sulfur-containing aerosol particles and H_2S from microbiological decomposition). Of the anthropogenic emissions, 93% are localized to the Northern Hemisphere and only 7% to the Southern Hemisphere, clearly indicating its industrial origin, namely, energy production from fossil fuels. Energy activities are responsible for about 90% of anthropogenic SO_2 and NO_x emissions. For SO_2, the main contributors are the stationary power sources, whereas for NO_x, the transport sector is equally important.

Technologies for the reduction of these emissions are already available and in use, but they do, of course, add to the production cost and sometimes decrease the net efficiency of the energy production process. The International Energy Agency (IEA)[9] has recently estimated that capital costs for flue gas desulfurization and selective catalytic reduction of NO_x account for 15 and 6%, respectively, of the cost of a new base-load coal facility. Typical reduction efficiencies are given in Tables 4 and 5. A general agreement to limit the emissions of SO_2 and NO_x from stationary power plants of larger capacity than 50 MW thermal has been passed within the European Community[11] (Table 6).

It is gratifying to note that these efforts have not been without result. According to forecast scenarios, SO_2 emissions will decrease between 1980 and 2010 to about 30% of their original value (but only a 10% reduction for the power production sector) (Figure 7) and the NO_x emissions by a substantially lower figure 20% due to the fact that satisfactory technical solutions do not yet exist, mainly for the diesel engines of the transport sector (Figure 8).

A substance of rather recent concern, for the reduction of which there is no satisfactory technical solution as yet, is CO_2 which, together with other so-called greenhouse gases such as CH_4 and N_2O, is present in the flue gases from fossil fuel combustion. The CO_2 content in the atmosphere is estimated to be responsible for over half of the contribution to the climate change (Table 7). Most of the anthropogenic CO_2 emissions are due to the burning of fossil fuel. The burning of biomass or other organic carbon is part of the organic carbon cycle and does not contribute to the net increase of the atmospheric CO_2 content. In addition to anthropogenic CO_2 emissions, the natural carbon cycle seems to be in imbalance, as more biomass (forest) is disappearing than is synthesized, resulting in a net contribution of CO_2 to the atmosphere.

Table 4
Typical Figures for Calcium-Based Desulfurization Processes

Method	Alkaline	Alkaline utilization (%)	Need for alkaline (kg/kg S)	Power consumption (kW/MW)	Amount of waste[a] (kg/kg S)
Wet scrubbing	$CaCO_3$	80–90	40[b]	4–7	5
Spray drying					
Without recirculation	CaO	70–80	2.5[b]	2.5–3.5	4.5
With recirculation	CaO	80–90	2.2[b]	3–4	4
Circulating fluidized bed combustion	$CaCO_3$	40–60	5.8–8.7[b]	3[c]	6–8
Injection method	$CaCO_3$, CaO, $Ca(OH)_2$	10–20	15–35d	0.5–0.1	7–10
Combined methods (Lifac)	$CaCO_3$	40–60	6.5–8.5	0.5–1.5	6–8

[a] Without ash in waste.
[b] 90% purity.
[c] Power consumption increase compared to PF firing without desulfurization.
[d] 90% $CaCO_3$.

From "Report on a Study Tour of the ECE/UNDP Inter-Country Project on International Cooperative Research on Low-Calorie-Value (LCV) Solid Fuel Technology", Technical Research Centre of Finland, Otaniemi, 1988. With permission.

Unlike the other emission components cited above, CO_2 emissions are not expected to decrease in the forseeable future unless drastic measures are taken. On the contrary, increasing economic activities in the developing world will add to the pressure (Figure 9).

The most immediate actions to curb this alarming trend must be policy actions. Within the European Community, it has been estimated that a decrease below the 1987 level could occur by the year 2010, with an annual increase in energy efficiency of 2.2% combined with an increase in the use of natural gas (from 13.7 Mtoe to 100 Mtoe by 2010) and nuclear energy (from 32.93 to 291 Mtoe) to meet the increased electricity demand. The rapid increase in nuclear energy seems a tough target to achieve despite the fact that the share of nuclear energy in the EEC electricity production has more than doubled between 1973 and 1983.

An increase in the energy efficiency of electricity production from fossil fuels combined with a subsequent shift in the types of fuel used, from coal to natural gas and later possibly toward increased use of biomass, could be the first steps in a strategy aimed at halting the

Table 5
Typical NO_x Emission Factors

Fuel	mg NO_x/MJ
Natural gas (100% load)	
Gas turbine	300
Gas turbine with steam or water injection	150
Gas turbine, premixed combustion	60–100
Condensing power plant	260
Industrial boiler	150
District heating central	50–100
Small scale boilers	30–50
Residual fuel oil	
High N	150–300
Low N	100–200
Distillate fuel oil	
Small scale	50–70
Coal firing	
Vortex burners, existing boilers	450
Diffusion burners and a large	
furnace in an existing boiler	230
Low-NO_x burners	140
Low-NO_x burners plus catalytic reduction	60
Fluidized bed combustion	115
Grate firing	150
Peat firing (existing boilers)	
Vortex burners	200–250
Vortex burners	300
Diffusion burners	200–400
Grate firing	200
Fluidized bed combustion	100–150
Peat firing (with low-NOx pulverized firing [PF])	130–160
Wood	60
Black liquor	50–80
Wood waste	
Grate	80–100

From "Report on a Study Tour of the ECE/UNDP Inter-Country Project on International Cooperative Research on Low-Calorie-Value (LCV) Solid Fuel Technolgy", Technical Research Centre of Finland, Otaniemi, 1988. With permission.

present alarming trends caused by excessive use of fossil fuels for electricity production.

Integrated combined cycle gasification processes (ICCG) (Figure 10) offer potential for a substantial increase of electricity production from thermal power plants. Although the advantages of the concept as such

Table 6
European Commission Emission Limit Values for New Plants Burning Solid Fuels

	Conditions	Limit value (mg/N/m^3)
Dust	Thermal capacity (MW)	
	≥500	50
	<500	100
NO$_x$	Solid in general	650
	Solid with less than 10% volatile compounds	1300
SO$_x$	Thermal capacity (MW)	
	≥500	400
	100–500	2000–400 (interpolation)

From "Report on a Study Tour of the ECE/UNDP Inter-Country Project on International Cooperative Research on Low-Calorie-Value (LCV) Solid Fuel Technolgy", Technical Research Centre of Finland, Otaniemi, 1988. With permission.

have been known for some time, recent technological developments, mainly on the material side for high-temperature gas turbines, have made them an attractive development alternative today. The main advantage of the processes, apart from the fact that a higher operating temperature gives a better thermodynamic efficiency for a heat engine (Section 2), is the possibility of increasing the proportion of electricity produced from the fuel, from 0.4 in a conventional condensing power plant up to 0.6 in an ICCG unit. As the trend in the increasing energy demand is toward a higher proportion of electricity use, it is important to develop technical solutions that can respond to this with the smallest possible increase in the total demand of primary energy.

Employing an ICCG process instead of a conventional coal-fired power plant results in much improved energy production efficiency, as seen from Figure 11, together with reduced emissions (Figure 12). If natural gas could be substituted for coal, the situation would further improve on the emission side, notably for CO_2, through increased efficiency (Figure 11).

Estimates have even been made of the cost of CO_2 removal from a fossil fuel-fired power plant. The technology for the removal of the CO_2 from the flue gases is, in fact, readily available (Figure 13); however, no satisfactory solution exists for the final deposition of the CO_2 removed.[13] Temporary solutions have been suggested, and even practiced

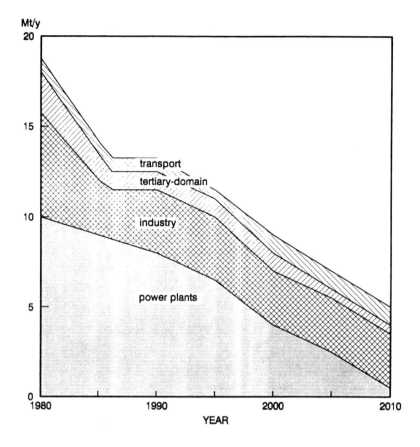

Figure 7. SO$_2$ emissions for EUR12, by sector, representing "conventional wisdom" approach using best available current technology. (Source: *Energy and the Environment.* COM (89) 369, Communication from the Commission to the Council, Commission of the European Communities, Brussels, February 8, 1990.)

at a limited level,[5] in which the compressed pure CO$_2$ gas is injected into gas or oil wells to enhance the recovery. Deposition in used salt mines has also been proposed. Although perhaps ecologically acceptable, these solution are of limited interest, as they cannot account for all the CO$_2$ produced, and are not available in the vicinity of all power plants. Suggestions of pumping the CO$_2$ in liquified form into the sea fulfill the capacity requirement, but involve ecological complications of an unknown nature.

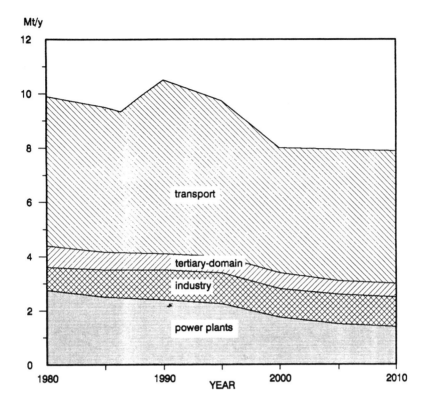

Figure 8. NO$_x$ emissions for the European community, by sector, representing "conventional wisdom" approach using best available current technology. (Source: *Energy and the Environment.* COM (89) 369, Communication from the Commission to the Council, Commission of the European Communities, Brussels, February 8, 1990.)

While searching for the perfect solution, one must not overlook any possibility to reduce the extent of the problem by immediate actions, temporary as they may be.

It is interesting to note that the cost of such an elaborate exercise as CO$_2$ removal from flue gases, while high, is not totally beyond the reach of present-day society; estimates indicate a 1.5 to twofold increase in the current electricity price.

Table 7
Greenhouse Gases and Contribution to Climate Change

Greenhouse gas	Concentration in atmosphere ppm	Contribution to climatic change %	Residence time in Atmosphere (a)
Carbon dioxide (CO_2)	350	45–55	Stable[a]
Freons (CFC gases)	0.0005	20–30	8–400
Methane (CH_4)	1.65	10–20	10
Nitrous oxide (N_2O)	0.3	5–0	170
Proposphoric ozone (O_3)	0.02[b]	5–10	About 1 d
Others		2–5	

[a] Normal constituent of the atmosphere (as water vapor).
[b] Occasionally up to 0.2 to 0.4 ppm in polluted air.

From Report of the Finnish Academies of Technology, Helsinki, Finland, 1990:1. With permission.

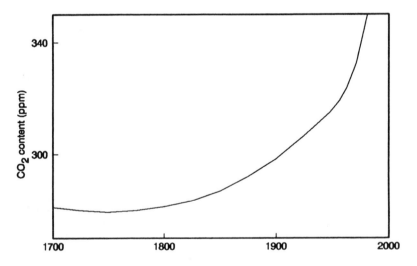

Figure 9. The evolution of atmospheric CO_2 content since 1700. Unless the emissions are curbed, the CO concentration will rise to 550 to 700 ppm by the year 2000. (From Report of the Finnish Academies of Technology, Helsinki, Finland, 1990:1. With permission.)

4.3.1. Fuel Cells

A fuel cell is a device which allows the conversion of the chemical energy in a fuel directly into electricity, without going through the traditional heat engine cycle. Thus, a fuel cell is not limited in its efficiency by the Carnot relation, and much higher conversion efficiencies

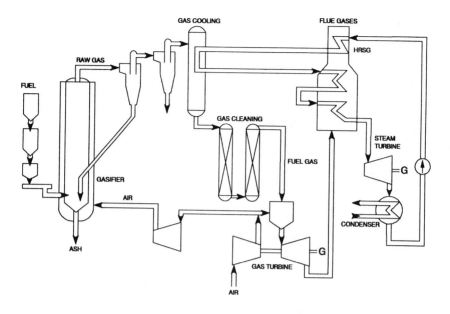

Figure 10. Schematic representation of a modern integrated gasification combined cycle (IGCC) power production process.

(60% and above) can be achieved. Basically, the fuel cell exploits the difference in chemical potential between two electrodes and converts this into an electric potential difference. When the circuit is closed, electrons flow through an outer circuit. The chemical potential difference is maintained by adding fuel equal to that consumed. The principle could be used to exploit the free chemical energy of many reactions, but in terms of clean engineering, the reaction between hydrogen and oxygen to form water is the reaction of preference, as only harmless water is produced as an emission. Hydrogen, of course, has to be produced either by splitting water (requiring electricity) or from other conventional fuels such as natural gas, fossil fuels, or biomass, creating other emissions (at least carbon dioxide if a carbon-containing fuel is used). Nevertheless, fuel cells hold great promise for clean and efficient electricity production for mobile use as well as large-scale power generation.[33]

A fuel cell can be viewed as a battery with a continuous supply of feed or as a reverse electrolysis. In the electrolysis of water, hydrogen and oxygen are formed at the cathode and anode, respectively. In a fuel cell,

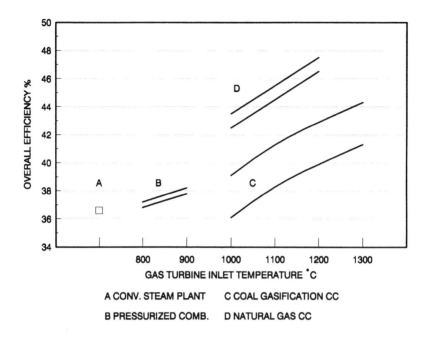

Figure 11. The efficiency of a power plant as a function of inlet gas turbine temperature.

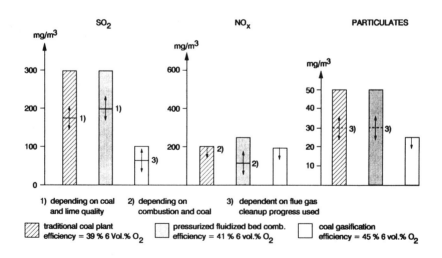

Figure 12. Emissions from coal-powered electricity plants of different technology. (Source: *Energ, wirtsch. Tagesfr.*, 3, 37, 1987.)

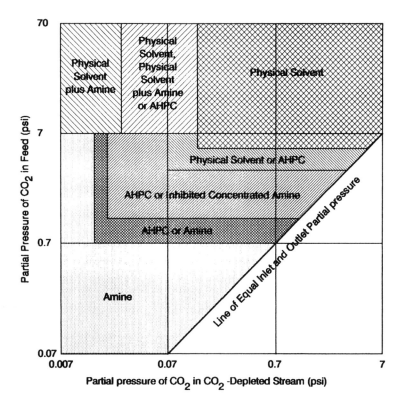

Figure 13. CO_2 removal from gases. (After Sparrow, F. T. et al., *Carbon Dioxide from the Gases for Entrained Oil Recovery*, ANL/CNSV-65, Argonne Laboratories, 1988.)

hydrogen and oxygen are added to the anode and cathode, whereby, through electrochemical reactions at the electrode surfaces, an electric current is created. As mentioned, fuels other than hydrogen could be oxidized in a fuel cell, resulting in the corresponding oxidation products as emissions. Contamination of the electrolyte in the fuel cell is a technical reason, other than environmental considerations, that makes hydrogen the fuel of preference for most fuel cell applications.

The theoretical voltage of a fuel cell (or an electrochemical cell in general) can be calculated from the Gibbs standard free energy of the overall reaction, provided the reactants and products are in their standard state.

$$\Delta G^0 = -nFV^0$$

where F is the Faraday constant and n the number of electrons transferred in the reaction. If the components are not in their standard state, the general form of the Nernst equation must be used:

$$V_{ideal} = V^0 + \frac{RT}{nF} \ln \frac{a_1}{a_2}$$

where a_1 = reactant activity and a_2 = product activity. The maximum thermal efficiency is the ratio between the free energy ΔG^0 of the combustion reaction of the fuel divided by the total energy content (enthalpy ΔH^0) of the fuel:

$$\eta_{ideal} = \Delta G^0 / \Delta H^0$$

The ideal efficiency for the reaction in question is less than 100% and decreases strongly with temperature if the change in entropy of the reaction is <0.

In real conditions, the ideal efficiency cannot be reached, since the voltage will be diminished due to polarization in the electrolyte at the electrodes as well as through the internal resistance of the cell.

$$V_{real} = V_{ideal} - v_{cathode} - v_{anode} - R_i I$$

where R_i is the internal resistance and v the polarization.

In general, a single cell produces an output of less than 1 V, and useful voltages are obtained by connecting many individual cells in a stack. Various types of fuel cells have been developed with different characteristics. The characteristics of the most important types are summarized in Table 8.

The alkaline fuel cell (AFC) is characterized by a high current density and provides high power production per unit volume. This is the reason for its application in space programs. From the point of view of wider application, the necessity of using absolutely CO_2-free hydrogen and air is a drawback.

Table 8
Characteristics of Various Fuel Cell Types[33]

	AFC	PAFC	SPFC	MCFC	SOFC
Temperature (°C)	60–120	150–250	80–100	650	900–1100
Anode fuel	H_2	H_2	H_2	H_2, CO	CH_2
Cathode fuel	O_2	Air	Air	Air + CO_2	Air
Electrolyte	KOH	H_3PO_4	Polymer	(K, Li) CO_3	Y_2O_3, ZrO_2
Anode catalyst	Pt	Pt	Pt	Ni	Ni/Zr_2O_3
Cathode catalyst	Pt	Pt	Pt	NiO	La-Sr-MnO_3

The phosphoric acid fuel cell (PAFC) is closest to commercial larger-scale utilization. It is not sensitive to carbon dioxide and can therefore be run on hydrogen produced from hydrocarbons. Care must, however, be taken to convert all CO, which is a poison for the anode reaction, to hydrogen through the shift reaction.

A number of companies have developed PAFC technology to commercial scale. A 4.8-MW unit and two 1-MW units were operated in Japan during the 1980s. An 11-MW plant will be installed and put on stream in 1991.[33] Other large demonstration units are a 1-MW unit to be installed in Milano in 1991 and a 5-MW unit in Japan in 1993. Although the PAFC technology cannot compete with more advanced fuel cell systems in electrical efficiency, it is compact and environmentally benign. These considerations make it an alternative for the extension of power supplies in densely populated or heavily polluted areas. The installation of 10- to 20-MW units at existing transformer stations is being considered in Tokyo. PAFC units may also have a market in decentralized power production in the 50- to 200-kW range.

The solid polymer fuel cell (SPFC) is based on solid polymeric materials, which limits its application to low temperatures. Its main application could be in the automobile market.

The molten carbonate fuel cell (MCFC) and the solid oxide fuel cell (SOFC) operate at high temperatures, which makes it possible to convert the hydrocarbon fuel inside the cell itself, resulting in a potential for very high efficiencies.

Of the developing technologies, MCFC technology is closest to commercial power generation. Units in the 20-kW range are operated in laboratories, and the first demonstration unit was started in Denmark in 1990. The American Public Power Association is planning a 2-MW

MCFC installation to be completed around 1993. A 1-MW unit is planned for mid-1990 in Japan, and several units in the 100- to 250-kW range are planned in the U.S., Japan, and Europe. SOFC technology is still in the development stage, but intensive research is being carried out in the U.S., Japan, and several countries in Europe.

4.4.   RENEWABLE ENERGY

Interest in the development of new renewable energy technologies comes not only from the need to diversify primary energy supplies, but also from the necessity of reducing emissions, primarily to air from fossil energy power generation. Today, the contribution of renewable energy is still rather small, on the order of 3% in the European Commnity and U.S., and around 15% for the whole world (mainly biomass utilization).[5] On the other hand, it is not insignificant, as manifested by the fact that the total amount of energy from renewable energy sources in the U.S. is larger than the combined energy produced by all of the nuclear power plants in the U.S.[13] Also, the projected economy looks encouraging (Table 9).

It is difficult to estimate the future potential of renewable energy resources, as the concept covers a great variety of different technologies. This variety is, of course, also an asset when trying to find an alternative with a broader base than the present fossil fuel-based energy economy. In any case, the introduction of new technologies will be slow; it is estimated that in the most favorable case, the total contribution of renewable energy sources — biomass, solar energy, hydro electricity, geothermal and wind energy cannot exceed 8% of the total energy consumption in the European Community by the year 2010.[5]

All energy flows are, strictly speaking, of solar origin, but when we speak of renewable energy production, one generally implies the use of solar radiation for energy production purposes.

For the purposes of energy production, the sun can be viewed as a black-body radiator (temperature of 5777 K) with a spectral distribution following Planck's law (Figure 14). The intensity of the solar radiation at a plane normal to the radiation outside the earth's atmosphere averages 1353 W/m^2. Interference with the atmosphere reduces the intensity and distorts the spectral distribution somewhat, depending on the optical path traversed by the radiation. When the sun is directly overhead (at the

Table 9
Cost of Electricity in Favorable Locations

Energy source	Present (1991)	Intermediate (1995–2000)	Long term (beyond 2000)
Photovoltaics	30–40	10–20	6
Wind	7–9	5	3.5
Biomass	5	5	4
Solar thermal	10	8	6–8
Geothermal	5–7	5–7	≤6

Note: Cost in cents per kilowatt hour.

Source: EPRI J., June 17, 1991.

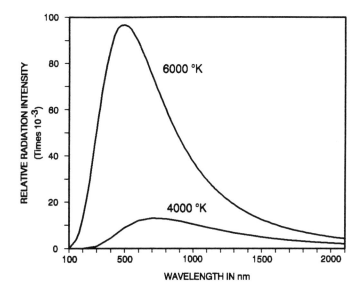

Figure 14. Radiation intensity at different wavelengths of a black body. The area under the curve between two wavelengths corresponds to the energy in Joules per second radiated from 1 cm² of a black body in the range of wavelengths.

equator), the optical path through the atmosphere is the shortest possible, which in technical terms is referred to as a situation of Air Mass 1 (AM1). The AM factor increases with higher (or lower) latitudes; at 45°, it is 1.41, and at 30° (60°), 2. For reference purposes, the value 1.5 (AM1.5) is generally used (Figure 15).[14] This radiation falling on the earth is the ultimate energy source for all activity on earth. In terms of the utilization of solar radiation for energy purposes, two basic options

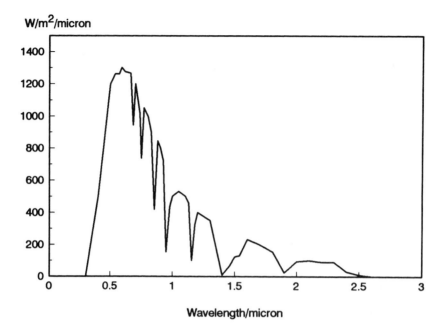

Figure 15. The spectrum of solar radiation at air mass 1.5. (From Spiers, D. J., *Solar Electricity and Solar Fuels,* NEMO Rep. 7, Technical Research Centre of Finland, 1989. With permission.)

can be followed by nature or man: that of converting the incident radiation to heat, or the more advanced form in which the rather unique properties of the flux of high-energy photons offered by such a high-temperature radiator are utilized. Most of the dynamics of the air and water masses on earth are a result of direct, but uneven, heat adsorption in the biosphere. All natural chemical phenomena are the result of a solar-driven reaction in which the photon flux plays a central role. Photosynthesis, through which biomass is synthesized, is the prime example of this. The use of solar radiation as a source of thermal energy involves adsorption of the incident radiation and conversion of the thermal energy for useful purposes.

The use of high-energy photon flux for energy production involves a first step in which charges are separated, with subsequent (simultaneous) transportation of the separated charges in opposite directions for utilization either for electricity production (photovoltaics) or for chemical reactions (photochemistry). These two basic options, already utilized in nature, are the only ones also available to man in his effort to use solar radiation as a source for renewable energy production purposes.

Direct conversion of solar radiation for energy purposes, although technically feasible, is subject to certain basic limitations. The most serious of these is the dispersity of the energy flow and its variation according to geographical or local conditions, as already mentioned; another is the fact that much of the radiation intensity reaching the earth's surface has been "downgraded" by scattering through the atmosphere, and here is no longer representative of the photon flux from the original high-temperature radiation source. The efficiency by which useful energy can be extracted from the incident sunlight depends on this dilution factor and varies according to the technology used (Figure 16).[15]

4.4.1. Biomass

Natural photosynthesis-producing biomass is the grand example of solar-driven complex chemical processes. It is not, however, a very energy efficient process in nature. Only about half of the incident light falls within the energy domain (0.38 to 0.68 µm) that can be used by plants. Of this portion, about one fifth is lost by reflection from the leaves. A large fraction of the absorbed radiation energy (about 40% for most plants) is lost through reactions (photorespiration) which do not produce carbohydrates. Suppression of this unnecessary (as it appears) side reaction is currently a topic of intensive research.

Photosynthetic CO_2 fixation involves a complex reaction apparatus resulting in inevitable additional losses, and the final efficiency for converting the incident light energy to biomass rarely exceeds 2% (a theoretical limit for the conversion efficiency of 12 to 13% has been calculated).[16]

In addition to the losses inherent in the photosynthetic reaction itself, we must consider other inefficiencies as well in the biomass-to-fuel production chain. These are a result of incomplete harvesting, energy expenditures for cultivation and harvesting, and conversion efficiencies for final product formation.

The usable output product (in terms of calorific value per square meter) W can be formulated as[17]

$$W = \int_0^\infty \eta * G_\lambda \, d\lambda (1-r)(g-1) f/g$$

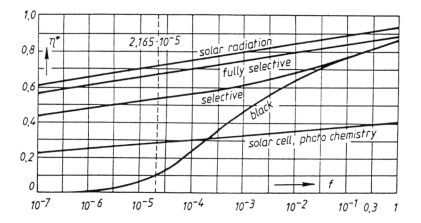

Figure 16. Energy content along various conversion paths in dependence on the dilution factor f of solar irradience $G = fs\,T_S^4$ assuming a Planck spectrum with $T_S = 5762$ K. (s = Stefan-Boltzmann radiation constant = $5.67 \cdot 10^{-8}$ Wm^{-2} K^{-4}.) The value f = 2.165 $\cdot\,10^{-5}$ corresponds to the irradiance of one solar constant, 1353 Wm^{-2}.[15]

where:

G_λ average spectral solar flux density (at wave length λ)

r   the fraction of biomass that remains in the fields as residue

g   energy harvest factor (ratio of the energy content of the biomass to the energy expenditure for growing the biomass, planting, fertilization, irrigation, harvesting, transport, etc.)

f   conversion efficiency of the raw biomass to the final product (e.g., ethanol or other liquid fuels)

η^* photosynthetic efficiency for biomass produced in the plant

$$\eta^* = \Delta H \Big/ \int_0^\infty \int_{t^*} G_\lambda \, d\lambda \, dt$$

where ΔH is the heat of combustion of dry biomass. Typical values are $\eta^* = 0.02$, r = 0.5, g = 2, and f = 0.8. All these values are time averages over a natural vegetation period t. With an isolation of

$$\int_0^\infty G_\lambda \, d\lambda = 140 \, \text{Wm}^{-2}$$

we have an average fuel production rate of $W = 0.6$ Wm^{-2}, equivalent to an energy efficiency of only 0.4%.

Different species of plants show different efficiencies (Figure 17), and improvement in natural photosynthesis efficiency appears to be most promising. In particular, recent progress in the genetic engineering of plants shows great promise as a method for improving the efficiency, particularly in suppressing (unnecessary) photorespiration.

Regardless of the potential for improving the natural photosynthetic process, the use of biomass as a renewable energy source presents the largest potential for immediate use. The amounts annually synthesized are roughly equivalent to ten times the world energy demand in calorific terms. The drawback is the same as for most renewable energy sources: it is dispersed and unevenly distributed, rarely found in great quantities where the energy consumption is largest.

Nevertheless, it has been estimated that the amount of biomass available for exploitation within the European Community, essentially in the form of agricultural and other waste, is on the order of 5% of the primary energy production, while forecasts show that the total potential including agricultural products for nonfood use, could correspond to about 5 to 10% of the EEC energy consumption before the year 2010, thus in the rather immediate future,[18] considering the generally long lead times in the energy production industry. According to one estimate,[19] biomass resources could replace all oil used in light-duty vehicles and all coal burned for power generation if vehicle fuel economy were doubled, and efficient gasification and combined cycle power generation units used.

In order to make better use of the potential of biomass-based energy production, a thorough systems study of the various opportunities should be done for different regions, as suggested by the IEA workshop on long-range R & D opportunities for renewable energy.[20] The great attractions for industrial exploitation of biomass are the large quantities available and flexibility in the choice of processes and end products (also other than energy). Bioconversion is a low-temperature process involving no hazardous chemicals, as opposed to the conventional chemical industry that often employs high temperatures and dangerous chemicals. Further, there is negligible augmentation of the greenhouse effect compared to the burning of fossil fuel, as biomass synthesis is

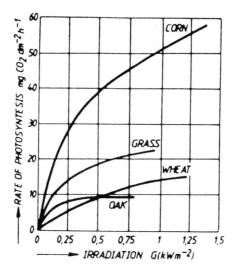

Figure 17. Role of photosynthesis as a function of irradiation level. (From Sizmann, R., Solar driven chemistry, *Chimia*, 43(7–8), 202, 1989. With permission.)

part of the natural carbon cycle (enhanced use may, however, cause temporary variations in CO_2 emissions). Up to now, the variety of local conditions, in terms of raw materials and product needs, has discouraged large-scale biomass utilization projects other than the traditional ones, pulp and paper and food industry, which are examples of large-scale operations based on renewable biomass resources. The flexibility in choice of raw material depending on need as well as production technology, should be an advantage if properly exploited.[20]

Crucial points to be considered before entering into large-scale operation are:

1. Choice of raw materials (plant species, plant breeding techniques for improved species, and other biomass sources (such as waste, etc.)

2. Plant cultivation methods

3. Long-term agricultural sustainability (need for fertilization, irrigation, possible crop rotation, and multispecies cultivation)

4. Raw material harvesting and gathering, transport, and storage

5. Pretreatment (mechanical, chemical)

6. Conversion to fuels (hydrogen, methanol, and ethanol), chemicals (ethylene, propylene, methanol, and benzene), and fine chemicals (agar, alginate precursors, and pharmaceuticals)

7. Downstream processes (separation and cleaning)

8. Transport and storage

9. Long-term environmental questions

The need for research in all of these sectors could be described by a development need index (DNI), as exemplified in Reference 20 (Table 10).

Various studies, including the FAST[18] projects, suggest that an integrated and complementary system for the use of biomass (agro-chemo-energy complex, i.e., multipurpose farms, biofactory, biomass refineries, agro-refineries, etc.) could be a technically, socially, and economically feasible concept, with energy use being an important parameter, but not the only purpose of the activity. However, any such activity would have to be judged on the basis of a systems approach, looking at the total impact of the activity and not just sectorial improvements, i.e., avoid creating environmental problems due to excessive fertilization or adding to the greenhouse effect through increased anaerobic activity (methane production).

The production of clean automobile fuels is another important area of development for biomass use. The concept of alternative automobile fuels based on renewable resources such as biomass is not new. As a result of the first oil crisis, much research and development effort was devoted to this issue during the 1970s; at that time, however, the driving force was fear of an imminent oil shortage, not environmental concern. Today, biofuels are a hot topic as a possible answer to many severe pollution problems caused by traffic. The gravity of the situation is illustrated by the fact that according to the EPA, 101 cities in the U.S. with a cumulative population of 140 million persons are suffering from poor air quality due to traffic emission.[21] In the cities of California, the average share of pollution from traffic is 43% of the hydrocarbon emission (VOC), 57% for nitrogen oxides, and 82% for carbon monoxide, which explains the recent attention to reducing the emissions of this sector.

Table 10
Development Need Index (DNI) Derived from the Total Cost (A) and the Degree of Commercialization (B) of Different Steps in Methane Production from Kelp

	Seeding and harvesting conversion gas			
	Outplanting			Cleaning
A: % of total cost	26	44	18	12
B: % level of commercialization[a]	25	75	25	100
DNI (A/B)	1.04	0.56	0.72	0.12

[a] 100% = commercially available; 75% = pilot scale; 50% = demonstration scale; 25% = laboratory scale; 5% = concept.

From Indegaard, M., Johansson, A., and Crawford, B., Jr., *Chimia*, 43(7–8), 230, 1989. With permission.

Biofuels do indeed offer potential for improvement in many areas related to emissions (Table 11). The most common biofuel is ethanol produced from biomass, the others are vegetable oils and biogas. Ethanol is already produced and used as a fuel in large quantities (Table 12).

There are no severe technical obstacles to a substantial increase of ethanol use. Additions in moderate concentrations (up to 10%) have shown no negative effects, in higher concentrations, some difficulties related to materials and cold start have occurred, but these can certainly be overcome.

Vegetable oils can be produced from the seeds of oilplants such as raps (Table 11). This oil can be used as a fuel in diesel engines, either directly or after modification. Direct use requires modification of a standard car diesel engine to avoid coking and clogging of nozzles. Modified engines that are designed to run on pure vegetable oil are commercially available.[22]

The other approach requires esterification of the vegetable oil with monoalcohols such as methanol or ethanol instead of the trialcohol glycerol present in the natural substance. This transformation is a simple procedure carried out at moderate temperature (60 to 80°C) in the presence of an alkali catalyst such as sodium hydroxide. Use must, however, be found for the glycerol liberated in the process; also, it is doubtful whether in many countries vegetable oil can be produced in sufficient quantities to have a significant impact on the transport fuel sector.

The proposal to make lubricating oils from vegetable oils, in particular for two-stroke engines, seems more attractive in terms of the quantities

Table 11
Properties of Various Fuels

Property	Ethanol		Vegetable oils		Biogas		Gasoline	Diesel fuel
	Otto	Diesel	Oil	RME[a]	Otto	Diesel		
Octane number	109		—	—	>120		95–98	—
Cetane number	<8		38–51	48–54		<10	25	>45
Lower heating value (MJ/l) (gas -200 bar)	21.1		34.4	32.7		4.1	32.5	35.5
Flame point (°C)	+13		+320	+111	-170		-40	>+50
Power	+++	++	–	–	– – –	– –		
Efficiency	+++	+	–	=	++	– –		
Emissions								
Carbon monoxide	+++	– –	–	–	+++	– – –		
Hydrocarbons	+++	– –	=	++	+++	– – –		
Nitrogen oxides	++	+++	+	–	–	+		
Polyaromatic hydrocarbons	+++	+++	+++	+++	+++	+++		
Aldehydes	– – –	– –	+?	+	– –	– ?		
Soot, smoke	++	+++	+	++	+++	+++		
Mutagenicity	+++	+++	++	++	+++	+++		

Note: +, =, -: compared to gasoline/diesel fuel; ?, no information, valid as an estimate based on fuel composition.

a Raps oil methylester.

From Nilsson, D., Alternative Fordon i USA—utveckling och kommersialisering, Utlands rapport Fråns Sveriges tekniska attacheér, U.S., 1990.

Table 12
Major Users of Alternative Fuels

Country	Total	LPG	Ethanol	CNG	Synthetic gasoline	Methanol	Electricity
Brazil	110	—	110	—	—	—	—
Japan	79	79	—	—	—	—	—
U.S.	62	18	34	1	—	9	—
Italy	57	42	—	15	—	—	—
New Zealand	45	3	—	9	33	—	—
Holland	27	27	—	—	—	—	—
Europe	18	9	—	—	—	9	—
Canada	8	7	—	1	—	—	—
U.K.	2	—	—	—	—	—	2
Australia	2	2	—	—	—	—	—
All others	37	15	8	14	—	—	—
Total	447	202	152	40	33	18	2

Note: Use estimated in thousands of barrels per day of gasoline equivalent. For comparison purposes, world gasoline consumption is 15.7 million bbl/d and U.S. gasoline consumption is 6.8 million bbl/d. Ethanol and methanol estimates are based on fuel production data. All others are based on the simplified assumption that vehicles use the equivalent of 800 gal of gasoline per year.

From U.S. Department of Energy, *Assessment of Costs & Benefits of Flexible and Alternative Fuel Use in the US Transportation Sector Progress Report Two: The International Experience*, DOE/PE-0085, Washington, D.C., August 1988.

and costs involved, and is not without importance from an environmental point of view, as emissions from increased use of chain saws and snow scooters as well as boating on small lakes has become a threat to delicate areas of nature. Such biodegradable two-stroke oils are already commercially available and, in fact, recommended for use on lakes in Germany and Switzerland.

Biogas is produced in anaerobic fermentation (rotting) of organic materials such as agricultural residues or urban refuse. Before use as a motor fuel, the methane in the gas must be separated from unwanted constituents such as carbon dioxide and hydrogen sulfide (as well as other sulfur-containing components). The purified gas can be used directly in Otto engines without modifications, although better economy can be attained through adjustment of the engine to the fuel. It can also be used in diesel-type engines either as an additive in the combustion air replacing part of the liquid fuel or as the main fuel after engine modification.

The use of gas as diesel fuel offers great potential for reducing emissions, notably particulate emissions. Technically, storage and refuelling present practical problems for which solutions have been suggested but not tested on a larger scale.

It is difficult to estimate the economy of biofuels compared to traditional petrochemistry, as it involves penetration into a very complex (and sensitive) area of agricultural policy in most countries. At least in order of magnitude, however, biofuels can be produced technically at prices comparable to their petrochemical counterparts. In terms of production economy, biomass-derived ethanol and methanol can already compete with coal-derived equivalents in small-scale fabrication, and ethanol is projected to be cost competitive with gasoline in the U.S. by the year 2000.[19] In these estimates, no penalty for emission has been included, which would no doubt further improve the competitivity of biofuels.

4.4.2. Photovoltaics

The photovoltaic effect is quantum mechanical in its origin, but in simplified terms it can be described as a process in which electrons of a particular semiconducting material, through the absorption of photons in the incident light, are excited to so-called higher-conducting energy states. The two charge carriers, the negatively charged electrons and the positive "electron holes" formed in the semiconductor, are separated by the potential difference created by the junction to another material, resulting in the ability to sustain an electric current under the influence of a photon flux of sufficient energy and intensity (Figure 18).

In theory, the conversion of radiation energy to electricity can proceed with high efficiency; the device is not limited by thermodynamic restrictions such as the Carnot efficiency valid for solar thermal installations. In practice, however, there are other inherent losses which limit the efficiencies of solar photovoltaic cells.

First, the excitation of electrons to the conduction band requires a minimum energy quantum, putting a lower limit on the photon energy that can be used. Photons of less energy are lost as heat. Photons of higher energy can excite electrons, but the energy in excess of the necessary excitation energy is again lost as dissipated heat. Further, some of the formed electron-hole pairs may recombine before the charges are separated, again contributing nothing but heat to the process. As the potential difference in a junction is rather small, several elements

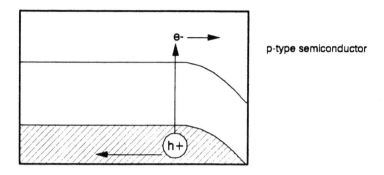

p-type semiconductor

Figure 18. Schematic illustration of the energy band of a p-type semiconductor.

are connected together to form a module. To provide more power, several modules may be connected together to form arrays.

There are two basic designs of photovoltaics: single junction cells and multi junction cells. The single junction cell can only be optimized for a small portion of the solar spectrum, determined by the choice of the material. The multi junction cell consists of several layers of semiconductors, active in different parts of the spectrum, stacked upon each other to yield a better total efficiency in terms of absorbed and converted sunlight. Typical efficiencies are given in Figure 19. The ultimate efficiency limit of a photovoltaic appears to be 40% for a tandem cell compared to 30% for a single cell.[15]

The average intensity of sunlight falling on the earth is clearly very dependent on the prevailing geographical situation and meteorological conditions. Even in the most favorable case the maximal energy per square meter is on the order of 0.3 kW,[23] which, of course, gives the ultimate limit for the power to be derived from a solar device (of any kind). Given the losses of a photovoltaic cell, the electricity produced is of the order of 120 W/m^2 in the best case, using expensive multilayer cells. To obtain more power, large areas of photovoltaics have to be used, or alternatively, the sunlight from large areas must be concentrated on a cell.

Both alternatives carry a penalty in cost. Today, photovoltaics (PV) are used in a variety of applications from satellite power generation modules of a few milliwatts to utility units as large as ARCO Solar's 6.6-MW power station in California. The price of large-scale electricity

Conversion Efficiency %

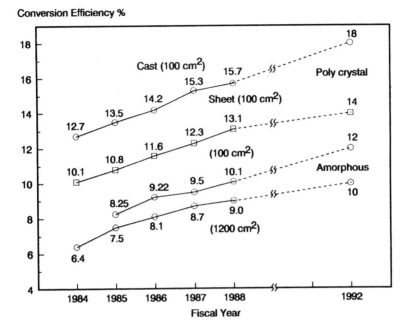

Figure 19. Conversion efficiencies of solar cells. (After Europe-Japan: The Global Environmental Technology Seminar Jetro, Tokyo, 1990.)

generation by PV is estimated to about 25 to 30¢/kWh — still high compared to the current tariff price, but a long way from the $10/kWh it used to be in 1975. The U.S. National Photovoltaics Program has set a short-term goal of 12¢/kWh and a long-term goal of 6¢/kWh. In order to reach these goals, several technical hurdles have to be passed. Current energy costs from photovoltaic systems are still 40 to 45¢/kWh, about three times greater than the price needed to make them an alternative for utility peak power applications. The growth of the photovoltaic market during 1974 to 1988 is shown in Figure 20, together with the decrease in the cost of peak power.

4.4.3. Solar Thermal

In solar thermal installations, the thermal energy of the incident solar radiation is utilized. They typically consist of three major elements: concentrators, receivers, and converters. The concentrators are optical devices such as mirrors or lenses that concentrate the sunlight from

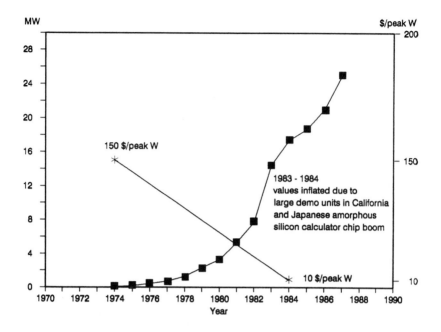

Figure 20. Worldwide shipments of photovoltaics (PV) estimated in peak megawatts.[14]

large areas into a receiver, which consists of a material with high absortivity and low emittivity, in which the incident photon energy is converted into heat in a transport medium. The heat is transported to a converter that converts the thermal energy into work or electricity (unless the device is used for heating purposes only).

Several types of concentrators are in use; some of the most common ones are described below, together with their concentration factors.

The low concentrators are typically used to reach temperatures of 100 to 400°C to generate process heat or steam for electricity generation, while the high concentrators, in which temperatures on the order of 3000°C can be reached, offer the potential to utilize the unique nature of concentrated photon fluxes for high-temperature photochemical reactions.

The cost of electricity now generated by solar thermal power plants is around 12¢/kWh when used in a cogeneration configuration in which the waste heat can be utilized. The target in U.S. programs is to lower the cost to 5¢/kWh, which involves fundamental improvements in the concentrator materials to cut costs as well as simplifications in the receiver systems.

Efforts have also been made to convert solar thermal energy directly to electricity through so-called thermionic converters. A thermionic converter consists of two electrodes, a heated cathode, and a cold anode. When heated (typically to temperatures of 1300 to 2800 K), the cathode emits electrons from its surface which are trapped at the cooler surface, creating a potential difference between the electrodes. If the electrodes are connected by an external resistance, current can be made to flow through the circuit driven by the temperature difference. Thermionic devices yielding up to 20 kW have been constructed.[24]

In principle, use of the thermoelectric effect (or Seebeck effect, after its discoverer T. Seebeck) could also be used for converting heat into electricity. The effect is based on the fact that a potential difference is created between two junctions of different materials that are kept at a different temperature. (The same effect is used for the measurement of temperature differences in a thermocouple.) The technique is limited to small power generation due to the very small potential differences created (on the order of 10^{-4} V/K) and relatively low efficiencies (~11%).

4.4.4. Wind Energy

Temperature differences at the earth's surface create pressure gradients in the surrounding atmosphere. Although extremely complex and impossible to calculate in detail, the movements of air masses due to these pressure gradients show some consistency and regularity locally over time, and this has throughout history been used by man. The moving gas masses carry kinetic energy (of the order of 0.2 kW/m^2) that can be harnessed by technological devices, such as sails, windmills and now wind machines for electricity generation.

A typical windmill contains four basic elements: (1) a rotor for capturing the kinetic energy of the moving air, (2) gearbox and transmission, (3) tower or fundament, and (4) generator and control electronics. The most vulnerable part of the windmill is the rotor device, which is sensitive to imbalances caused by rain or snow and ice.

Windmills vary greatly in size, from a few kilowatts to more than 4 MW. The average size today is 50 to 100 kW, but is expected to rise to 100 to 250 kW with the accumulated experience of the present stations.

Presently, in California there is about 2 GW of installed power based on wind energy, and it is expected to grow by 300 to 400 MW annually.[25] The cost of energy produced is 8 to 12 ¢/kWh, depending on the location and size of the unit. The goal of the U.S. program is to lower the cost to 4 ¢/kWh.

Wind is generated through the combined effects of pressure gradients in the atmosphere and the Coriolis force caused by the rotation of the earth. Wind streams that are not influenced by friction from the earth's surface are called geostrophic winds.

Depending on the stability of the atmospheric conditions, the altitude at which the winds can be said to be geostrophic ranges between 100 m and 1 km. The average speed of the geostrophic winds is, in a way, the maximum average windspeeds attainable in a certain region. Figure 21 shows the geographical variations in wind speeds over Northern Europe.

At altitudes of interest for wind power generation, wind speeds are affected by atmospheric stability, surface vegetation, topography, and single objects such as large buildings. The influence of atmospheric conditions on wind speed is schematically pictured in Figure 22.

In neutral conditions, the average profile of wind speeds as a function of altitude can be described by the formula:

$$v(Z) = v(Z_1)\left(\ln\left(\frac{Z}{Z_0}\right) \Big/ \ln\left(\frac{Z_1}{Z_0}\right) \right) \tag{1}$$

where:

v (Z)   wind speed at altitude Z
v (Z$_1$)   wind speed at altitude Z$_1$
Z$_0$ roughness of the surface in wind direction

Large deviations from the average can occur due to gusts; the turbulence is strongest in nonstable atmospheric conditions and over rough surfaces.

The distribution of wind speed is reasonably well described by the Weibull distribution function[26] (Figure 23).

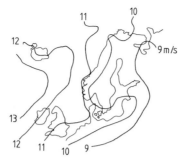

Figure 21. Geographic wind speeds (average) in Northern Europe. (From Peltola, E. NEMO Rep 9, Technical Research Centre of Finland, 1989. With permission.)

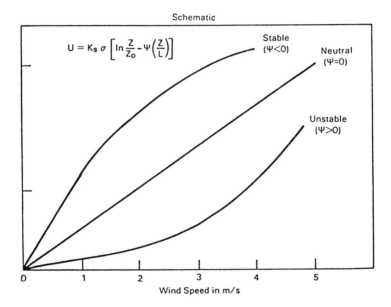

Figure 22. Influence of atmospheric conditions on wind speed. (From Peltola, E., NEMO Rep. 9, Technical Research Centre of Finland, 1989. With permission.)

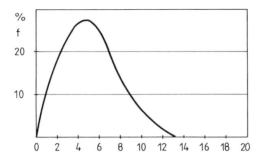

Figure 23. Distribution of wind speed. (From Peltola, E., NEMO Rep. 9, Technical Research Centre of Finland, 1989. With permission.)

$$F\left(v < v_x\right) = 1 - e^{\left(v_x/c\right)^k}$$

$$f(v) = \left(\frac{k}{c}\right)\left(\frac{v}{c}\right)^{k-1} e^{-\left(v/c\right)^k} \qquad (2)$$

where:

c   parameter fixing the location of the distribution (m/s) function
k   parameter fixing the form of the distribution
F(v) cumulative wind speed distribution function
f(v) frequency function

The form of the distribution function influences the energy content of the wind in such a way that the energy content increases as k decreases. Small k signifies the frequent occurrence of wind speeds higher than the average.

Constant variations in wind speed are a result of periodic and stochastic variations. Both effects are of importance for wind energy production, as the periodic variations in energy production must be considered in relation to the energy demand, and the stochastic variations influence the need for storage or back-up power generation capacity.

Information on these variations must be gathered from statistical meteorological data for the particular sites considered. From these data, the average annual potential for wind energy production can be estimated for given regions.

$$e_r = C_{tot} \cdot e_t = C_{tot} \cdot T(a_1 - a_2 \log(z_R)) \tag{3}$$

where:

z_R roughness index for the whole area, composed of a weighted average for the different terrain conditions present

T annual hours (8760 h)

a_1, a_2 constants that relate the roughness of the surface and prevailing wind speeds to energy potential

e_t average annual wind energy per surface area

C_{tot} average energy efficiency of wind generators ($\approx 30\%$)

e_r energy potential/rotor sweep area

Typical values for a_1 and a_2 for the estimation of wind energy potential are $a_1 = 210$ W/m^2 and $a_2 = 160$ W/m^2 for h = 100 m, and $a_1 = 120$ W/m^2 and $a_2 = 130$ W/m^2 for h = 50 m.

These types of calculations can be made to roughly estimate the potential for wind energy production in larger geographical regions such as whole nations. Such a survey has been carried out at least for the EEC countries,[27] Finland,[26] and Sweden.[28] In addition to the potential for wind power generation, these surveys also considered the availability of land for wind energy production, taking into account alternative land use possibilities. More detailed calculations can also be made, taking into account the actual type and individual configuration of the windmills for optimal efficiency.[1]

The actual cost of power production by windmills can be estimated from the formula:

$$h = (C_1 + C_{OM})k_{ADD}H_T/e_r \tag{4}$$

where:

H_T cost of windmill at factory/rotor sweep area
k_{ADD} coefficient to include installation costs
C_1 annuity factor for investment cost
C_{OM} annual operating and maintenance cost in relation to investment
e_r estimated power production of windmill/rotor sweep area

Listed below are typical values of H_T (European prices), K_{ADD}, C_I, and C_{OM}.

H_T \$350/m^2 for 50h, \$650/m^2 for 100 h
k_{ADD} earth, 1.6; coastal region, 1.8; sea, 2.0
C_I 0.08 (= 8%) (5% interest, 20-year lifetime)
C_{OM} 0.02 (= 2%)

4.4.5. Future of Renewable Energy Production

On the whole, it can be said that renewable energy production methods have undergone remarkable progress during the last decades, not so much due to scientific breakthroughs as to progress in application and engineering. Many of these new technologies are projected to be competitive with fossil-based energy production in the relatively near future (Figure 24). The effect has no doubt been due to many long-term national energy research programs launched in several industrialized countries, which has brought about the determination and continuity necessary for technical progress in this area, for a long time dominated by the thought that only a centralized power generation and distribution system can be viable. It is probable that the current trend of incorporating environmental considerations into process economics will further accelerate this trend.

4.4.6. Fusion

Research on harnessing fusion energy for peaceful purposes began in the mid-1950s, with expectations of solving the world's energy problems within the next 20 years. The fundamental problems of maintaining a fusion reaction (heating and confining the plasma), however, turned out to be much more difficult to solve than anticipated.

Despite the difficulties, the research has continued without interruption in the hope that a solution would provide a decisive improvement in the world energy situation, as the fuel used, deuterium, is an abundant element and the fusion process itself does not produce radioactive wastes. Further, the inherent safety of the fusion reaction, e.g., a failure in the cooling system automatically leads to extinction of the reaction,, is an obvious advantage compared to the fission reaction.

At this point, however, these advantages, although fundamental, are somewhat speculative, as they will finally depend on the technical

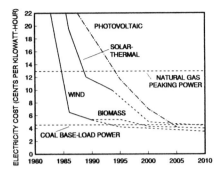 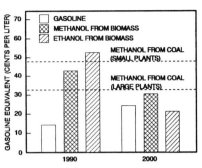

Figure 24. Biomass may well be a viable energy alternative in the near future in industrialized countries (above evaluation for U.S.)[19] (From Weinberrg, C. J. and Wiliams, H. R., *Sci. Am.*, 263, 3146, 1990. With permission.)

realization of the process, which has turned out to be far from straightforward. Already at this point, it appears clear that harnessing fusion energy will not be simple or cheap.

To initiate a fusion reaction, the temperature of the fuel has to be raised to some 100 million degrees at which point the kinetic energy of the totally ionized nuclei (about 10 to 20 keV) is sufficient to overcome the repulsive forces of the Coulomb barrier. At this temperature, the hot fuel in a plasma state must be confined to a given density in order to allow the fusion reaction to proceed sufficiently.

The product n^* τ^*T, where n^* = density of the fuel, τ^* = time of confinement, and T = temperature of the fuel, is a measure of fusion reaction conditions. These conditions are most easily achieved for a fuel consisting of a mixture of the heavy isotopes of hydrogen, deuterium (D), and tritium (T). A net energy production from a DT plasma is possible when n^* τ^* T $> 10^{21}$ m^{-3} s keV; this condition is known as the Lawson criterion.

The reaction products in a DT fusion are α-particles (helium nuclei) and neutrons in addition to an energy release of 17.6 MeV (corresponding to a fuel value of 93,500 Wh/g). The liberated α–particles carry 3.5 MeV of the released energy, and as they carry a coulombic charge, they will remain in the plasma and lose their kinetic energy to the plasma through collisions. This released energy is sufficient to maintain the temperature of the fusion reaction without external heating.

The requirement for plasma ignition is n^* τ^* T $> 5 \times 10^{21}$ m^{-3}/s keV, i.e., a fivefold harder condition than that for net energy production.[29]

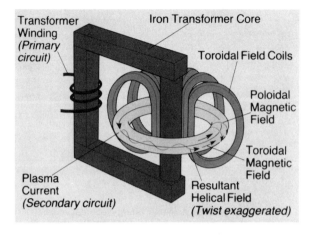

Figure 25. The plasma torus in a tokamak reactor. (From Joint European Torus Programme (JET), Oxfordshire, England. With permission.)

The introduction in the 1960s of the tokamak reactor developed in the Soviet Union was a major breakthrough in fusion research. The tokamak is a closed "magnetic bottle" in which the plasma is in the form of a torus, held together by a magnetic field generated by ring-shaped magnetic windings which enclose the torus (the field lines running in the direction of the large torus diameter) (Figure 25). By induction, a strong current is generated in the plasma circuit, which (in addition to heating the plasma) again generates a poloidal magnetic field around the plasma torus. This induced field is additive to the external toroidal field in a way that the field lines wind around the torus in a helical fashion.

As the current is induced (which requires a pulsed mode of operation), the tokamak cannot be run continuously. Typical cycles range from 100 to 6000 (5400 burn time, 450 dead time[30]). Presently, only burn times in the range of 1 to 10 s have been achieved, due to restrictions in the magnetic field. These restrictions are believed to be overcome by superconducting magnets. Figure 26[29] shows the results achieved to date in tokamak reactors.

The tokamak research work has advanced steadily within several international research programs and the next generation of reactors is at an advanced planning stage at both the European NET (Next European Torus) program as a direct continuation of the JET (Joint European Torus) program and the IAEA-initiated ITER (International Thermonuclear Experimental Reactor) program in which, in addition to

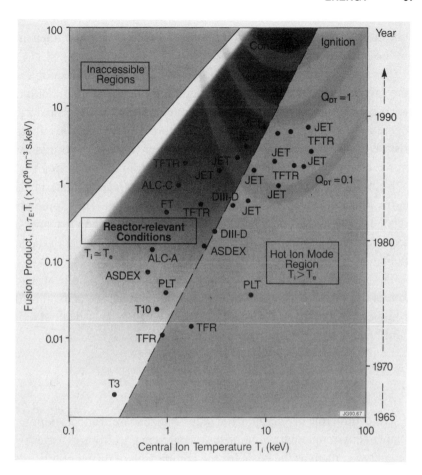

Figure 26. Achieved values of fusion product and ion temperature in various tokamak experiments. Note recent value achieved by the JET project (1989). (From Heikkinen, J., Karttunen, S., Alava, M., Pättikangas, T., and Salomaa, R., *The State of the Art and Future of Fusion Technology,* Rep. TKK-F-B129, Helsinki University of Technology. With permission from JET, Oxfordshire, England.)

European countries, Japan, the U.S., and the Soviet Union participate. As this phase is going to be very costly — the estimated investment for the ITER unit of gigawatt size is $3.5 to $4.9 billion, and the total projected cost of the research program over 20 years is on the order of $10 billion — international cooperation is clearly necessary.

It is hoped that this next phase will prove the technical and economic viability of fusion energy production. The central questions that remain to be resolved are problems related to fuel impurities as well as material questions related to the extreme temperatures and radiation intensities in the inner parts of the tokamak.

The inherent advantage of a fusion reaction is its passive safety, i.e., the fusion reactions are always extinguished in case of malfunction. It is also believed that the radiation from a fusion reactor in normal operation can be kept well below current norms. Nevertheless, radiation risks are involved. Possible emission sources are contaminants dissolved in the cooling system and tritium emissions from the fuel circuit.

Neutrons of the fusion reaction activate the reactor material and corrosion components dissolved in the cooling circuit. These problems, related to waste treatment and decommissioning of fusion reactors, are expected to be similar to, and of the same magnitude, as that of the low radioactive waste of fission reactors.

The current timetable of research and development indicates that unless major breakthroughs in technology or science occur, the first commercial application of tokamak-based fusion energy production could occur at the earliest by 2030 to 2040.

4.5.   NET ENERGY ANALYSIS

The central role of energy in the industrialized world and its dominant share of environmental pollution has inspired many efforts to use energy or energy content as the basis for the comparison of different processes or alternatives.[1,31]

Net energy analysis (NEA) is an unidimensional value assessment method in which all values are expressed in terms of energy content. NEA is a departure from multidimensional economic theories and does not include market-driven incentives. The basic assumption is that energy is the essential input to all production processes. All energy-requiring items associated with the energy investment are converted to energy units (materials, labor, transportation, maintenance, etc). These are combined into a single factor, the net energy ratio:

$$r = e_0 / e_i \tag{5}$$

where e_0 is the total energy produced by the system over its entire lifetime and e_i is the total energy needed for construction and maintenance of the system. To be a net energy producer, the energy output must exceed the energy investment over the lifetime, i.e., $r > 1$.

Some modifications must be incorporated into Equation 5 for an analysis of harmful emissions to the environment. We assume that the new installation capacity replacing the old technology does not produce emissions. For the denominator, we use total fossil fuel input, and the numerator is the amount of clean energy produced over the lifetime.

We may define a net nonfossil energy ratio:

$$R = E_0 / E_i \qquad (6)$$

where E_0 is the total emission-free energy production and E_i is the total fossil energy needed to construct the capacity in question. The total fossil fuel energy represents a mix of various fuels, as the commodities needed are not produced by one fuel only. Total emissions would then be

$$\text{total emissions of component } k = \sum_j f_j E_i I_{jk} \qquad (7)$$

where j is the fuel type, f_j is the fraction of fuel j of the total, and I_{jk} is the emission matrix per unit energy of type j for component k.

To begin with, a national or global fuel mix for f_j may be used for the fossil fuel input. Note that even if energy conservation, and renewable and nuclear energy installations may be inherently emission free, some fossil fuel (which will inevitably involve emissions) will still be needed to construct the facilities.

Compared to the starting point energy mix, the emissions will be inversely proportional to the net nonfossil energy ratio (or net energy ratio), which typically is of the order of 5 to 20,[32] which corresponds to an 80 to 95% reduction of emission units.

The forms of energy input greatly affect the R factor, i.e., thermal energy is considered less valuable by a factor corresponding to the Carnot efficiency. Alternatively, all values have to be compared to some reference primary fuel such as coal, e.g., in terms of coal equivalences. In Table 13, output energies are electricity except for solar heating.

For renewable energies local conditions may greatly influence the value of R — by a factor of two to three for solar energy between north

Table 13
Reported Net Energy Ratios (R)

Energy source	R^a
Nuclear fission	8–15
Nuclear fusion	7
Solar cells	4–8
Solar space power	5–18
Solar heating	5–10
Central solar power	10
Wind energy	8–13
OTEC (ocean thermal)	10
Wave energy	2

[a] Indicative figures only.

and south, and the yearly output of wind energy may drop by 50% if the average wind speed drops from 7 to 6 m/s.

4.5.1. Energy Breeders

The principle of the energy breeder concept is to assume that part or all of the constructed new clean-energy capacity is used as an energy investment for the new clean-energy capacity.

Initial construction of the first capacity will require an amount of fossil fuels q_0 over the time interval $-t_b-t_L/R$; $-t_L/R$ where t_b = building time and t_L = effective service time. At $t = 0$, the input fossil fuel energy has been paid back, there is now a nonfossil capacity of q_0, and the breeding phase can start, i.e., a certain fraction of the clean energy is directed into investments for further clean capacity.

As energy and electricity data show an exponential growth with time, it is reasonable to assume that the same kind of exponential growth trends are applicable to new technology options to reduce emissions. Thus, the total new installed capacity in the breeder phase can be written as

$$q(t) = q_0 \exp(t \ln 2/T) \tag{8}$$

where t = time, T = exponential doubling time (in years), and q_o = nonfossil new generation capacity at $t = 0$ or fossil energy needed to construct the first unit (J/year). The existing installed capacity now doubles for each T year.

With a construction time t_b (years) and an operational lifetime of t_L (years), the net nonfossil energy output at time t can be shown to follow the relation:

$$Q_{net}(t) = q(t) * \left\{ 1 - t_L \middle/ (Rt_b) * \left\{ \exp(t_b \ln 2/T) - 1 \right\} * \left\{ \exp(t_L \ln 2/T) + 1 \right\} \right\}$$

(9)

It is practical to define a net gain factor that is the ratio of the net nonfossil energy produced at time t to the capacity q_0 at time t = 0:

$$G(t) = Q_{net}(t)/q_0$$

(10)

G > 0 indicates net nonfossil energy production and is the prerequisite for energy breeding. The maximum penetration speed is obtained by setting G(t) = 0, which is fulfilled by $T = t_L/R$ and $t_b = T$. This corresponds to a situation where all net energy is directed to new energy investment and nothing is left for consumption.

From Equation 8 we have (Figure 27):

$$q(t) = q_0 \exp(Rt \ln 2/t_L) = q_0 2^{\left(\frac{Rt}{t_L} \right)}, \text{ Figure 27}$$

(11)

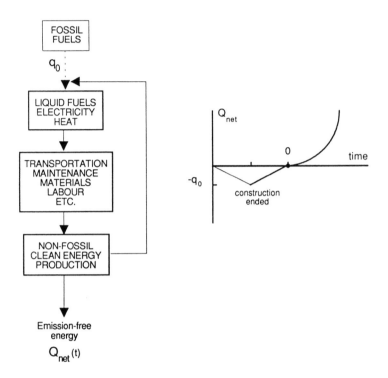

Figure 27. The energy breeder.(From Lund, P., in *IEA/OECD Expert Seminar on Energy Technologies for Reducing Emissions of Greenhouse Gases,* Organisation for Economic Cooperation and Development, Paris, 1989. With permission.)

REFERENCES:

1. Slesser, M. *Energy in the Economy,* Macmillan, New York, 1978, 7.
2. Gluckman, M. J. *Toward 21st Century Coal Processing*, Electric Power Research Institute 1991, World Coal Institute Conference, London, April 3–5, 1991.
3. Tolba, M. Children and the environment, in *The State of the Environment,* UNEP, United Nations Environmental Programme, 1990.
4. World Commission on Environment and Development, *Our Common Future,* Oxford University Press, New York, 1987.

5. *Energy and the Environment,* COM (89) 369, Communication from the Commission to the Council. Commission of the European Communities, Brussels, February 8, 1990.

6. Ausubel, J. H., Grüber, A., and Nacicenovic, N. Carbon dioxide emissions in a methane economy, *Climatic Change,* 12, 245, 1988.

7. Ohta, T. "The world needs to find a clean energy source", *The Japan Times,* November 16, 1990, 5.

8. Rogner, H.-H. and Britton, F. E. K. *Energy Growth and the Environment: Framework for Discussion,* Commission of the European Communities, Eco Plan International Report, Paris, 1991.

9. IEA, *Emission Controls,* International Energy Agency, Paris, 1988.

10. Commission of the European Communities, OJ L336, (88/609/EEC), Council Directive on the Limitation of Emissions of Certain Pollutants into the Air from Large Combustion Installations above 50 MW thermal, Brussels, December 7, 1988.

11. St. Clair, J. H. et al. Process to recover CO_2 from the gas gets first large-scale layout in Texas, *Oil Gas J.,* 7, 14, 1983.

12. Hepola, J., Solantausta, Y., and Johansson, A. *Recovery of Carbon Dioxide from the Flue Gas and Product Gases of Energy Production Processes,* Res. Note 1190, Technical Research Centre of Finland, 1990 (in Finnish).

13. Hubbard, H. M. Scientific and technical challenges in electricity generation, *Chimia,* 54 (7–8), 197, 1989.

14. Spiers, D. J. *Solar Electricity and Solar Fuels,* NEMO Rep. 7, Technical Research Centre of Finland, Otaniami, 1989.

15. Sizmann, R. Solar radiative energy, in *Proceedings of Summer School at Ingls, Austria, 31 July – 3 August 1985,* European Space Agency, ESA SP.240, 1986.

16. Bolton, J. R. Solar Fuels, *Science,* 4369, 705, 1978.

17. Sizmann, R. Solar driven chemistry, *Chimia,* 43(7–8), 202, 1989.

18. Forecast and Assessment in Science and Technology (FAST) Commission of the European Communities, EC Research Program, Brussels, 1984–1987.

19. Weinberg, C. J. and Williams, H. R. Energy from the sun, *Sci. Am.,* 263 (3)146, 1990.

20. Indergaard, M., Johansson, A., and Crawford, B., Jr. Biomass technologies, *Chimia,* 43(7–8), 230, 1989.

21. Nilsson, D. *Alternative fordon i USA — utveckling och kommersialisering,* Utlands rapport från Sveriges tekniska attacheér, Stockholm, 9009, 1990.

22. Elsbett engine in Germany, personal communication, 1990.

23. Minder, R., Wolf, M., and Leidner, J. R. The sun as a source of radiation, *Chimia,* 42, 124, 1988.

24. Grathwohl, M. *World Energy Supply,* Walter de Gruyter, Berlin, 1982, 262 (and references therein).

25. Lindley, D. *Large Scale Wind Power. A Review,* Symposium, Industry and Wind Energy, Dipoli, Finland, 1988.

26. Peltola, E. *The Potential of Wind Energy in Finland,* NEMO Rep. 9, Technical Research Centre of Finland, 1989 (in Finnish).

27. Selzer, H. *Wind Energy. Potential Wind Energy in the European Community. An Assessment Study,* D. Reidel, 1986.

28. Skärbäck, E. Potentialer för vindkraft, in IVA Symp., Stockholm, September 20, 1984.

29. Heikkinen, J., Karttunen, S., Alava, M. Pättikangas, T., and Salomaa, R. *The State of the Art and Future of Fusion Technology,* Rep. TKK-F-B129, Helsinki University of Technology, 1990.
30. Grathwohl, M. *World Energy Supply,* Walter de Gruyter, Berlin, 1982, 205.
31. Odum, H. T. and Odum, E. C. *Energy Basis for Man and Nature,* McGraw-Hill, New York, 1976.
32. Lund, P. Assessment of the effectiveness of renewable and advanced technologies in reducing greenhouse gases based on net energy analysis: the energy breeder concept, in *IEA/OECD Expert Seminar on Energy Technologies for Reducing Emissions of Greenhouse Gases,* Organisation for Economic Cooperation and Development, Paris, 1989.
33. Rostrup-Nielsen, J. R. and Christiansen, L. J. Fuel cell power plants for high efficiency: principles and status, in *Proceedings from 1991 Symposium on Energy and Environment,* ASHRAE, American Society of Heating, Refrigerating and Air-Conditioning Engineers, special publication, 52, 1991.

CHAPTER 5

Engineering

Engineering has long traditions — in fact, longer than those of science itself. In the beginning, it was rather an empirical "art" based on careful observations of natural processes, with occasional original inventions.

Today, science has become the mighty ally of all aspects of technology, to the extent that we can already speak about scientific technology rather than the traditional science and technology. The distinction is not only semantic; the scientific approach allows us to foresee events before carrying out the actual experiment; it also allows us to imagine details in the natural laws that could be put to use. Virtually the whole electronic industry is an example of an activity that exploits the knowledge of the fine points of quantum mechanics which would not have been likely to develop based on simple observation and trial and error.

We cannot change the laws of nature, but the task of engineering and technology (or scientific technology) is to find ways of circumventing them in order to direct things in a desired direction or avoid undesired effects. The full power of this machinery has not yet been used to avoid or abate the harmful environmental effects of technology. In fact, the methods we today apply to this respect are hopelessly clumsy in comparison with the technology which is at their origin, as evidenced by oil spills at sea or toxic waste handling (suffice to recall the almost tragicomic tales of ships that sail the seven seas like the haunted "Flying Dutchman", prohibited from entering any port with their cargo of toxic wastes).

One could say that the whole question of waste generation and waste handling is the real manifestation of the shortcomings of the present-day technological approach, and hence the need for a new approach where the potential negative effects of a particular technology or product are considered in advance, rather than corrected at great expense when the difficulties are encountered. Basically, this means that the concept of technology and engineering has to be broadened to encompass the whole life cycle of products and production. That is the meaning of the concept of clean technology or nonwaste technology.

5.1.   SEPARATION

Separation processes are central to almost all aspects of technology. Extraction of useful components from a bulk mass — atoms or molecules from each other, meaningful signals from noise, information from nonsense — are everyday problems of today's engineering.

In all cases up to now, the immediate concern has been focused on the product to be recovered, with little or no attention paid to the residue or waste left behind. Particularly in chemical engineering, it has become evident that such an approach is no longer satisfactory; perhaps the same problem will arise within other sectors such as electronics, as the useful transmission bands are crammed with noise.

Reduction of waste within the industry is to a great extent synonymous with increasing the efficiency of separation processes. By efficiency, we mean both the "sharpness" of the separation, i.e., how well the components are separated from each other, and the amount of energy required to effect the separation. The minimal energy required, defined by the entropy of mixing (section 2) two components, is given by:

$$\Delta S_{min} = -RT\left(x_1 \ln\left(a_1 x_1\right) + x_2 \ln\left(a_2 x_2\right)\right)$$

where a_1 and a_2 are factors accounting for the nonideality of the system.

From this, we immediately can deduce the discouraging fact that the concentration of an infinitely dilute solution requires an infinite amount of energy, or consequently, removal of the last atoms require huge amounts of energy, a fact of more than theoretical interest when we remember that a water stream containing a few parts per million of impurities may already be considered polluted.

In addition to the above energy of mixing (or actually demixing), one must add the difference in the energy of attraction between the components to be separated and that of the bulk phase. Depending on the size of this interaction energy, we speak about chemical reaction energy, solvation energy, or simple cohesion.

A general approach in most separation processes is that the system has to be brought to a state of equilibrium, at which point the components are separated from each other. Generally, this means that the attractive forces mentioned above have to be counteracted by some other effect, most often by changes in temperature. An example of this is distillation, the most common separation process, where the solution is heated to its boiling point, at which point the molecules, due to their high kinetic energy, move relatively freely with respect to each other in the vapor. This vapor, in equilibrium with the solution but of a different composition than the liquid phase, is then separated and condensed.

Another approach is to take the system very far from the equilibrium and then utilize the difference in the mobility of the different species as they race each other toward equilibrium. Processes of this kind usually take advantage of differences in diffusion coefficients, or in chemical or physical properties with respect to the external force by which the gradient deequilibrating the system is created. This force can be created by an electric field as in electrophoresis, chemical potential as in liquid extraction or similar processes, or simple gravity as in sedimentation or centrifugation. Differences in particle size can be utilized at the macroscopic level with sieves, or at the molecular level with membranes or zeolites (molecular sieves) (Figure 1).

In the chemical process industry, separation processes account for a large part of the investments and a significant portion of the total energy consumption. The dominant separation process in the chemical industry (for liquids) is distillation. In terms of clean engineering, an important aim is to find methods that provide a sharper separation than distillation, thus reducing the amounts of contaminated product streams (i.e., waste), improving the use of raw materials, and yielding better energy economy.

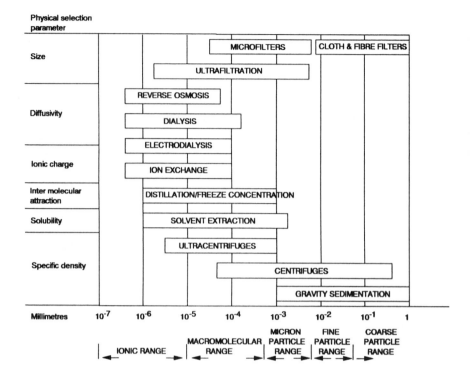

Figure 1. The relationship of various separation technologies to particle size. (Source Ben Aim, R. and Vladan, M., UNEP Ind. Env., 12(1), 15, 1989.)

In addition, the implementation of closed cycles (i.e., closed water cycle in the pulping industry) for the reduction of effluents frequently requires new separation techniques to allow the selective bleeding off of harmful or toxic compounds. It is for this purpose that some unconventional techniques offering potential for high separation efficiency and selectivity have been included in this section.

5.1.1. Supercritical Extraction

Supercritical extraction is essentially a liquid extraction process employing compressed gases under supercritical conditions instead of solvents. The extraction characteristics are based on the solvent properties of the compressed gases or mixtures.

Basic knowledge of the solvent power of supercritical gases or liquids has been known for more than 100 years,[1] but the first technical applications were not suggested until the 1960s, and the first industrial application commenced operation in the late 1970s.

From an environmental point of view, the choice of extraction gas is critical, and in the author's view, to date, only the use of CO_2 would qualify as an environmentally benign solution. Supercritical extraction with CO_2 has been applied in different branches of the chemical process industry (Table 1).

From a chemical engineering point of view, the advantage offered by supercritical extraction is that it combines the positive properties of both gases and liquids, i.e., low viscosity with high density, which results in good transport properties and high solvent capacity, in addition to the fact that under supercritical conditions, solvent characteristics can be varied over a wide range by means of pressure and temperature changes (Figure 2).

Relevant pressure ranges are from 8 to 50 MPa. The solvent properties of a gas can be drastically changed by adding another component to the solvent gas; mixing organic components into CO_2 generally enhances their solvent power while inert gases (Ar, N_2) reduce the solvent power, drastically. The latter effect could be used to further improve the energetics of a supercritical extraction process provided the extraction gas could be regenerated in an efficient way by, e.g., a membrane separation process.[2] The rapid development in the field today may make supercritical extraction an alternative worth considering not only for fine chemical separation, but also for bulk processes.

5.1.2. Membranes

Membrane processes constitute a well-established branch of separation techniques (Table 2). The particular interest in membranes in modern separation processes is that they work on continuous flows, are easily automated, and can be adapted to work on several physical parameters, such as:

- Molecular size
- Ionic character of compounds
- Polarity
- Hydrophilic/hydrophobic character of components

Table 1
Examples of Commercial Application of Supercritical Extraction of Natural Products

Active Components in Pharmaceuticals and Cosmetics	
Ginger	Calmus
Camomile	Carrots
Marigold	Rosemary
Thyme	Salvia

Spices and aromas for food	
Basil	Gardamon
Coriander	Ginger
Lovage root	Marjoram
Vanilla	Myristica
Paprika	Pepper

Odoriferous Substances for Perfumes	
Angelica root	Ginger
Peach and orange leafs	Parsley seed
Vanilla	Vetiver
Oil of spices	

Aromas for Drink	
Angelica root	Ginger
Calamus	Juniper

Further Applications	
Separation of pesticides	
Refinement of raw extract material	
Separation of liquids	
Extraction of cholesterol	

From Saari, M., *Prosessiteollisuuden Erotusmenetelmät*, VTT Res. Note and references therein, 1987, 730. With permission.

Microfiltration, ultrafiltration, and reverse osmosis differ mainly in the size of the particles that can be separated by the membrane. Microfiltration uses membranes having pore diameters of 0.1 to 10 μm. These can be used to filter suspended particles, bacteria, or large colloids from solutions. Ultrafiltration uses membranes having pore diameters in the range of 22,000 Å. These can be used to filter dissolved macromolecules such as proteins from solutions. Reverse osmosis membranes have pores so small that they are in the range of the thermal motion of polymer chains, i.e., 5 to 20 Å. Electrodialysis membranes separate ions from aqueous solutions under the driving force of an electrostatic potential difference. Membrane separation processes typically require low energy input, the

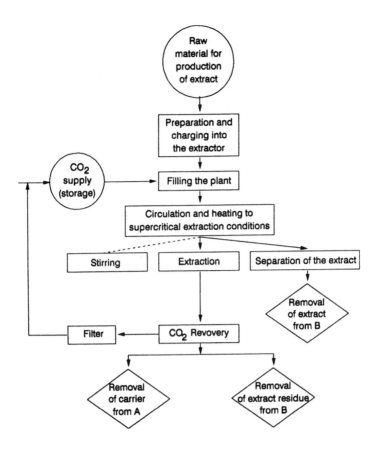

Figure 2. Flowchart of an extraction process using supercritical CO_2.

major operating expenditure coming from membrane replacement (Table 3).

5.1.3. Reverse Osmosis

Reverse osmosis is generally based on the use of membranes that are permeable only to the solvent component, which in most applications is water. The osmotic pressure due to the concentration gradient between the solutions on both sides of the membrane has to be counteracted by an external pressure applied on the side of the concentrate in order to create a solvent flux through the membrane. Desalting of water is one area where reverse osmosis already is an established technique.

Table 2
Main Membrane Separation Processes: Operating Principles and Application

Separation process	Membrane type	Driving force	Method of separation	Range of application
Microfiltration	Symmetric micro-porous membrane, 0.1 to 10 µA pore radius	Hydrostatic pressure difference, 0.1 to 1 bar	Sieving mechanism due to pore radius and adsorption	Sterile filtration clarification
Ultrafiltration	Asymmetric microporous membrane, 1 to 10 µA pore radius	Hydrostatic presssure difference, 0.5 to 5 bar	Sieving mechanism	Separation of macro-molecular solutions
Reverse osmosis	Asymmetric "skin-type" membrane	Hydrostatic pressure, 20 to 100 bar	Solution-diffusion mechanism	Separation of salt and micro-solutes from solutions
Dialysis	Symmetric micro-porous membrane, 0.1 to 10 µA pore size	Concentration gradient	Diffusion in convection-free layer	Separation of salts and microsolutes from macro-molecular solutions
Electro-dialysis	Cation and anion exchange membranes	Electrical potential gradient	Electrical charge of particle and size	Desalting of ionic şolution
Gas separation	Homogenous or porous polymer	Hydrostatic pressure concentration gradient	Solubility, diffusion	Separation from gas mixture
Supported liquid membranes	Symmetric micro-porous membrane with adsorbed organic liquid	Chemical gradient	Solution diffusion via carrier	Separation
Membrane distillation	Microporous membrane	Vapor-pressure	Vapor transport into hydrophobic membrane	Ultrapure water concentration of solutions

From Orioli, E., Molinari, R., Calabrio, V., and Gasile, A. B., in Membrane Technology for Production — Integrated Pollution Control Systems, Seminar on the Role of the Chemical Industry in Environmental Protection, CHEM/SEM.18/R.19, Geneva, 1989.

Table 3
Capital and Operating Costs of a Membrane Process

		% of total
Capital cost		
(including 5 year mortgage at 15% interest rate)	0.362	
Operating cost		
Maintenance (cleaning and general, 3% of initial cost)	0.036	15.4
Labor (0.2 man years at 20,000 ECU/year)	0.006	2.6
Energy (0.083 ECU/kWh)	0.091	38.5
Membrane replacement (every 2 years)	0.103	43.5
Total (operating)	0.236	100
Total (capital and operating)	0.598	

Note: Service factor, 82%; cost in ECU/m^3 feed on 1991 basis.

From Gaeta, S. N., et al., in New Technologies for the Rational Use of Energy in the Textile Industry in Europe, CEC Directorate XVII, Thermie Programme, Milan, October 16 to 18, 1991. With permission.

At the beginning of 1988, 6250 desalting plants were operating world-wide, with a total capacity of 3 billion gal/d. It is estimated that in the U.S. today, there are about 7050 desalting plants with individual capacities of more than 25,000 gal/d, or a total capacity of about 200 million gal/d. Reverse osmosis accounts for about 75% of this capacity, and 70% of the plants are for industrial use. Reverse osmosis is the most energy efficient process of desalination available today (Table 4). In addition, with the development of better membranes, capital and operating costs have been reduced by operation at lower pressures. Energy requirements for brackish water plants operating at 400 and 250 psig are 7.6 and 3 to 5.7 kWh/1000 gal, respectively. Capital operating costs for brackish-water desalination range from $0.60 to $1.60/gal/d plant capacity and $1.00 to $1.25/1000 gal of products, respectively.

The energy requirement for seawater desalination operated at 800 to 1000 psig is 27 kWh/1000 gal. Capital and operating costs for seawater desalination plants range from $4.00 to $10.00/gal/d plant capacity and $2.50 to $4.00/1000 gallons of product, respectively.[3] For seawater desalination, there is no need to improve the present stability and reliability of the membranes themselves, but improvement is needed in membrane selectivity, oxidative stability, and fouling characteristics.

Table 4

Energy Requirement for Water Provision

	Technology involved	Approximate energy requirement/ 1000 m^3 (GJ)
Run-off water in mountainous area	Dam	0.2–2.0[a]
Purified run-off water	Reservoir and plant	2.2[a]
Groundwater, 100 m deep	Pump	4[a]
Transport in pipelines (1000 miles)	Pipelines	60[a]
Transport in ships (1000 miles)	Super tankers	75[a]
Seawater distillation	Flash evaporation	280[a]
	Reverse osmosis	2.4[b]
		22[c]

[a] Slesser, M., *Energy in the Economy*, Macmillan, New York, 1978, 7.
[b] Brackish water.
[c] Seawater.

The major field for future work will be lowering the operating pressure currently required in desalination by reverse osmosis.

5.1.4. Ultrafiltration

Ultrafiltration is used for the separation of two components of different molecular mass. The size of the membrane pores constitutes the sieve mesh covering a range on the order of 0.002 to 0.05 µm. The permeability of the membrane to the solvent is generally quite high, which may cause an accumulation of the raffinate phase (to be concentrated) close to the surface of the membrane, resulting in increased filtration resistance, i.e., membrane polarization and back diffusion.

5.1.5. Electrodialysis

Electrodialysis is used for the separation of ionic components in an electric field in the presence of semipermeable membranes permeable only to anions or cations. Consecutive alternating membranes form a package of parallel cells (Figure 3).

If a saline solution is circulated through the package, the positive ions are drawn to the cathode and the negative ions to the anode by the effect of the electric field applied perpendicularly to the membrane surfaces. The net effect is a concentration of salts in every other compartment and a dilution of salts in the other compartments. As mentioned previously,

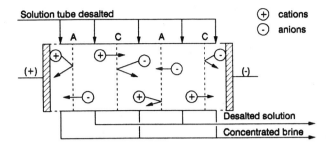

Figure 3. Schematic presentation of the operation principle of electrodialysis.

the reverse of the same physical phenomenon has been proposed for electricity production. Applications are desalination of water or recuperation of ionic components such as hydrofluoric acid. Examples of some technical realizations of membrane models are shown in Figure 4.

5.1.6. Pervaporization

In pervaporation, a liquid mixture is presented to the surface of the membrane and the permeate is removed as vapor from the downstream side. The mass flux is maintained by sustaining the vapor pressure on the permeate side of the membrane at a lower level than the partial pressure of the liquid feed. The usual method of operation is to sustain the low pressure of the permeate side of the membrane by condensing the permeate vapor.

The technique permits the fractionation of liquid mixtures by partial vaporization through a membrane, one side of which is under reduced pressure or flushed by a gas stream. At present, the only industrial application of pervaporation is the dehydration of organic solvents, particularly dehydration of 90%-plus ethanol solutions, although other applications are under development.[4] Depending on the membrane used, the result may vary from that obtained by simple distillation, e.g., permitting (in some cases) the separation of azeotropes.

5.1.7. Liquid Membranes

Liquid membrane technology offers a novel membrane separation method in which the separation is effected by the solubility of the

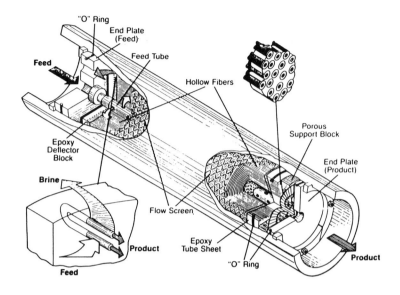

Figure 4. Sketch of a Permasep hollow-fiber reverse osmosis module. (Courtesy of Du Pont Co.)

component to be separated into a liquid membrane rather than by its permeation through pores, as is the case in conventional membrane processes such as ultrafiltration and reverse osmosis. The component to be separated is extracted from the continuous phase to the surface of the liquid membrane, through which it diffuses into the interior liquid phase.

The liquid membrane can be created in an emulsion or on a stabilizing surface of a permeable support (polymer film, glass, clay, etc.; Figure 5). The advantage of an emulsion is the rather large specific surface area. A simplified flow diagram of a continuous emulsion liquid extraction process is presented in Figure 6.

The emulsion is prepared in the first stage of the process (water in oil [W/O] emulsion) by emulgation of the inner liquid phase I in an organic phase II. In the permeation stage, the emulsion is dispersed into the continuous phase to be treated to form a water/oil/water (W/O/W) emulsion in which the actual material transfer takes place. In the sedimentation stage, the continuous phase is separated from the emulsion by gravity. In the fourth stage, the W/O emulsion is broken and the organic inner phase is separated.

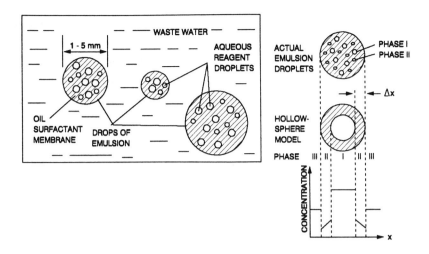

Figure 5. Liquid membrane drops. (From Saari, M., *Prosessiteollisuuden Erotus menetelmat,* VTT Res. Note (and references therein), 1987, 730. With permission.)

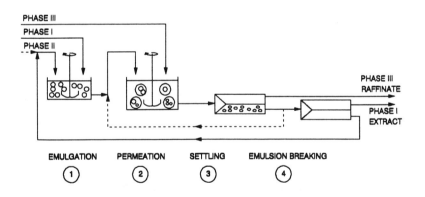

Figure 6. Continuous emulsion liquid membrane extraction. (From Marr, R., Lackner, H., and Draxler, J., in *VTT Symposium 102.* Vol. 1, 1989, 345. With permission.)

Although liquid membrane extraction is not yet widely available, promising results have been reported for a variety of applications, and it is claimed to offer distinct advantages over alternative methods.[5] The advantages most relevant for clean production technology compared to biological treatment are:

1. It is rather insensitive to process disturbances or changes in component concentration or flow speed.

2. It functions in a broader temperature interval than biological processes, which are generally limited to a narrow interval of 20 to 40°C.

3. It requires little space.

4. It does not produce waste sludge that creates a secondary environmental problem.

5. Impurity components may be recovered as useful chemicals, i.e., phenols in contaminated wastewater, instead of being destroyed as in a biological process.

A disadvantage is the fact that the organic phase is in direct contact with the continuous phase to be treated, with the risk that small amounts of organic phase material are carried away with the raffinate. This can be minimized by careful choice of the organic emulgating chemical (i.e., paraffinic rather than aromatic compounds, etc.).

Compared to ion exchange, the advantages are:

1. Liquid membranes are not as easily contaminated as resins.

2. The capacity is large and large material flows can be treated.

3. It is a continuous process.

4. It involves less materials, as high concentrations can be attained in the inner phase, and the material requirements for the membrane phase are small.

5. Its selectivity for many contaminants is better than that of ion exchange.

Compared to liquid extraction, liquid membrane extraction:

1. Requires less solvent

2. Is more compact, as the extraction and stripping are performed simultaneously

Cost estimates based on pilot plant trials for some metal extraction processes (Cu, V) indicate that liquid membrane extraction compares favorably to conventional liquid extraction processes[5] (Table 5).

Currently, the main areas of development are increasing the emulsion stability and the possibility of including catalyzed reactions in the inner phase.

Table 5
Cost of Cu and V Extraction[5]

	Cu		V	
	Liquid membrane	Liquid-liquid	Liquid membrane	Liquid-liquid
Recovered metal, (tons/year)	36,000	36,000	72	72
Extraction stages	1	5	1	4
Investment (millions of DM)	25.1	40.8	0.47	0.86
Operating cost (DM/kg)	0.11	0.12	10	16

Note: DM, Deutsche mark.

Recently, a large-scale plant utilizing liquid membrane technology was built in the Austrian viscose and rayon industry by Lenzig AG to resolve wastewater problems.[6] The wastewater comes from the spin bath and contains about 500 mg/l of Zn, which is separated selectively from calcium down to less than 3 ppm. Several other processes such as precipitation with H_2S and $Ca(OH)_2$, ion exchange resins, and solvent extraction had been tested, but the liquid membrane technique proved to be the most economic solution. The design and final choice of process was based on a 2-year pilot plant trial optimizing the process parameters for the liquid membrane process.

Figure 7 shows the schematic flowchart of the process with mass flows and concentrations. The mass transfer apparatus used is a countercurrent extraction column with a diameter of 1.6 m and an active height of 10 m. The emulsion is prepared in a homogenizer, and the emulsion splitting is done in an electrostatic emulsion splitter. After the screening of several extractants, bis(2- ethylhexyl)dithiophosphoric acid (DTPA) proved to be the most satisfactory. The technical and economic feasibility of the process was proved during the first months of operation, yielding an operation cost of 8.16 Austrian shillings (1/12 U.S. $) per kilogram of zinc recovered (Table 6).

5.1.8. Adsorption

Adsorption is a result of the fact that the intermolecular attraction forces between the molecules constituting the bulk phase or dissolved in it (gas or liquid) are weaker than the attractive forces between these molecules and those of a solid surface. This leads to an enrichment or adsorption layer of molecules from the bulk phase on the "attractive"

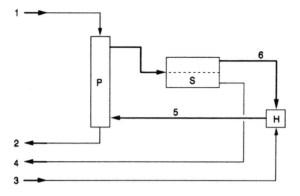

Figure 7. Flowchart of a plant for zinc recovery.

1 Wastewater, 75 m³/h, 500 mg/l Zn
2 Raffinate, 75 m³/h, 1 to 3 mg/l Zn
3 Stripping phase, 0.5 m³/h, 250 g/l H_2SO_4
4 Stripping phase, 0.5 m³/h, 60 g/l Zn
5 Emulsion, 5.5 m³/h
6 Organic membrane phase, 5 m³/h
P Permeation column
S Emulsion splitter
H Homogenizer

(From Marr, R., Lackner, H., and Draxler, J., in *VTT Symposium 102,* Vol. 1, 1989, 345. With permission.)

Table 6
Operating Costs of Liquid Membrane Technology (Austrian Shillings)

H_2SO_4	2.25
Neutralization	2.10
Electric energy	0.84
Organic loss	1.16
Evaporation	1.81
Total	8.16

surface. The magnitude of the interaction, which determines the efficiency of the adsorption, clearly depends on the molecular properties of both the solid surface and the surrounding bulk phase.

Adsorption is a well-established industrial unit operation employing a variety of materials for different purposes. Discovery of the peculiar properties of a group of substances named zeolites (some 200 years after discovery of the natural zeolites[7]), often descriptively called "molecular sieves", is considered a major breakthrough in the area of adsorption technology.

Zeolites are a class of crystalline compounds found in nature possessing cation exchange properties similar to other clay or kaolin-type aluminosilicates such as permutit.[8] The chemical composition of zeolites can be written as:

$$M_{\frac{2}{n}}OAl_2O_3 \cdot xSiO_2 \cdot y \cdot H_2O$$

where n is the cation valency, x is ≥ 2, and y is a function of the porosity of the framework.[9] A peculiar property of zeolites is that they can lose water continuously without suffering destruction of their crystal structure. Water can make up as much as half of the bulk density of zeolites, and evaporation of water leaves pores of varying size which can be used for (selective) adsorption of molecules.

Although the ion exchange properties of natural zeolites have been known and utilized for a long time, it was not until first synthesized in the late 1950s[10,11] that the full potential of tailor-made zeolites as molecular sieves was realized. Today, the synthesis and evaluation of the properties of new structures, and possibly those of related materials such as lantanide phosphates, is a thriving field of research (both for catalytic and separative purposes).

The selective adsorption of molecules on zeolites is due to the properties of the crystal structure. The size and shape of the network determines the size of the molecules adsorbed.

The cations of the zeolite influence the molecular sieve properties and selectivity of the zeolites through their charge, atomic radius, density, and position in the lattice. The selectivity can be influenced within certain limits by changing the cations, and may be of importance in tuning the selectivity for particular compounds from a solution containing several similar substances (Figure 8).

The present applications of zeolites fall into two categories: purification and separation. The purification properties of zeolites are generally due to their selectivity toward polar or polarizable molecules, while the separation properties mainly derive from their molecular sieve properties. Some typical adsorption applications are shown in Table 7. A promising potential application in environmental technology is in the purification of contaminated, dilute waste streams.

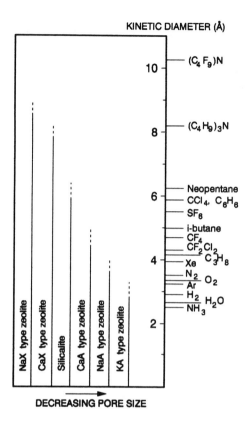

Figure 8. Correlation between effective pore size of zeolites in adsorption at 77 to 420 K, as determined from the Lennard-Jones force constants. (After Ruthven, D. M., *Chem. Eng. Progr.*, February, 42, 1988.)

5.1.9. Parametric Pumping

Parametric pumping is not really a separation method, but, rather, a trick by which the separation efficiency in some methods can be greatly increased. The principle of parametric pumping is to provoke periodic perturbations to the process equilibrium and flows in order to enhance the separation of liquid or gas components. Separation methods such as adsorption extraction or ion chromatography are, in principle, suitable for parametric pumping in cases where the intensive variables of the system such as temperature, pressure, pH, or electric field can be easily changed to effectuate an incremental, reversible change in the distribution of the components between the moving and immobilized phases. The

Table 7
Purification with Zeolite Adsorbents

Feed	G/L*	Adsorbent	Process Details
Drying of			
Natural gas	G	4A	T-swing
Air	G	4A	T-swing (also P = swing)
Refrigerants (chlorocarbons)	G/L	Modified 4A	Non-regenerative or T-swing
Cracked gas	G	3A	T-swing
Organic solvents	L	3A	T-swing
Acid gas	G	Chabazite	T-swing
CO_2 removal from air in submarines, spacecraft	G	4A	P-swing (vacuum)
H_2S removal from sour gas	G	CaA or Ca chabazite	T-swing
SO_x, NO_x removal (Purasiv) from air	G	Silicate	T-swing
Kr^{85} removal from air	G	Silicate or de-alum. H-mordenite	Chromatographic
I^{129} removal from air	G	AgX, Ag mordenite	T-swing
Concentration of alcohols from dilute aqueous alcohol	L	3A, 4A	T-swing

From Saari, M., *Prosessiteollisuuden Erotosmenetelmät,* VTT Res. Note (and references therein), 1987, 730. With permission.

principle is illustrated by the schematic of the equipment used by Wilhelm, the originator of the method[12] (Figure 9).

Reversal of the flow direction is effected by the pistons at each end of the column, and the separation is due mainly to temperature differences in the adsorption layer. The liquid to be treated is pumped through a heat exchanger in which its temperature is elevated. The heated liquid warms up the adsorption layer, thereby decreasing the adsorption capacity, and the excess of the adsorbate moves into the liquid phase. Cooler liquid is introduced into the upper part of the column, increasing the adsorption capacity in this part. Thus, the cyclic temperature variations of the

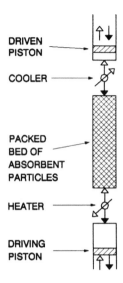

DRIVEN PISTON

COOLER

PACKED BED OF ABSORBENT PARTICLES

HEATER

DRIVING PISTON

Figure 9. Parametric pumping, recuperative thermal technique.

adsorption layer provoke cyclic variations in the mass flows in the adsorption interface layer, resulting in variations in concentration. The product in this batch process moves to opposite ends of the column. The results of the first experiments were rather discouraging (giving a separation factor of only 1.5), and the real breakthrough of the method was due to the introduction of the direct heating technique[12] (Figure 10).

In this batch process, the separation factor could be raised to more than 100,000 for a solution of toluene and heptane (the separation factor is the final ratio of the concentrations in the upper and lower part of the column.

The direct heating approach has been successfully used to enhance the efficiency of liquid extraction processes,[13–16] ion exchange,[17–21] and separation of gases.[22–23] Enhanced separation effects can be achieved by creating periodic heating and cooling intervals in a column without reversal of the flow direction (zone adsorption).[5] Industrial applications have not been reported yet, but the method seems to offer good potential for improving purification processes.

5.1.10. Biosorbents

A number of selected nonliving, inactivated materials of biological origin, such as algae, fungi, bacteria, and their products, have been screened for their uptake of metals from solution.[14] It was noticed that

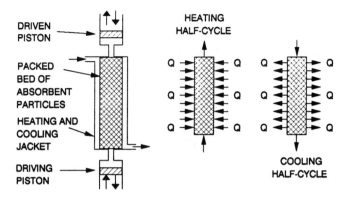

Figure 10. Parametric pumping with direct heating.

the different biomass types exhibited different performance for the metal species tested. The choice of pH for the test solutions also influenced biomass adsorption performance.

In particular, the biomass of the green alga *Halimeda opuntia* and the brown alga *Sargassum natans* have been identified for their good potential to concentrate chromium and gold, respectively. The optimum pH was 4 to 6 for chromium uptake and <3 for gold uptake. The brown marine alga *Ascophyllum nodosum* sequestered cobalt well at pH 4.5, outperforming activated carbon and ion exchange resins. The optimum pH for silver uptake by the biomass of the red alga *Chondris crispus* and for arsenic uptake by the yeast *Saccaromyces cerevisiae* covered a broad range of pH 2 to 6 and 4 to 9, respectively. The red marine alga *Palmaria palmata* was the most powerful biomass in sequestering platinum at a pH below 3.

The biosorptive uptake of chromium and gold was rather specific. It was not affected by the presence of other cations such as Cu^{2+}, UO^{2+}, Ca^{2+}, and Ni^{2+} or anions such as NO^-, SO^{2-}, and PO^{3-}. The kinetics of the metal uptake, except for gold, was generally very rapid.

Sequestered metals can be eluted from the biomass, and the biosorbent material subsequently reused many times over. The work is an early demonstration of the existing potential of a new group of biosorbent materials of microbial origin which can be effectively used in novel processes of metal recovery from dilute solutions.

5.2.   PROCESS DEVELOPMENT

5.2.1. Centralization/Decentralization/Integration

The question of large centralized production facilities vs. small decentralized ones is of prime importance when we consider alternative, more environmentally benign production methods. The answer is not obvious. It can be argued quite correctly that large centralized facilities allow, through economy of scale, larger possibilities of investing in efficient purification methods. On the other hand, it can equally well be said that many of the environmental problems we see today are results of excessive local exploitation of both raw materials and the surroundings as a whole. In fact, many of the waste-handling difficulties that we encounter today are due to the enormous quantities of waste produced, rather than the type of waste. It is often a problem of quantity rather than of quality.

The question of scale is especially pertinent when considering new renewable energy production methods. Generally, these rely on the utilization of dispersed primary energy sources, and have a difficult time competing with already established centralized power production and distribution systems. Introduction of the 1978 Public Utility Regulatory Act (PURPA) in the U.S. has improved theis situation considerably, and it is quite likely that a number of regulations will follow suit elsewhere in the world.

Intensive specialized farming is another example of how the quantity of a waste can become a problem if the local infrastructure is lacking. Previously farming was a truly nonwaste society, with most of the byproducts from one operation, by necessity, being used in another activity within the farm. With increased specialization and volume, there is no natural recipient for the residues, which become a burden rather than an asset as before.

There is one more lesson to be learned from this example: although today's farming residues are, in general, biocompatible and biodegradable, the surrounding environment cannot assimilate the quantities produced during modern intensive farming. A central question that remains to be answered is, are there any technical means of breaking the general rule for economy of scale in order to make small industrial operations economically feasible?

There are many examples of how an integrated energy production scheme can result in greatly improved energy economy for a process, and also enable the utilization of low-grade heat. Arguments have been advanced for carrying over the same type of thinking to raw-material streams, in the sense that the waste material from one process would serve as a raw material for another. While this might be a feasible strategy within a company, it is doubtful that it could be a sustainable strategy in the long run in a more general sense. Great caution must be exercised not to create a chain of interdependent processes subsidized in the interest of waste handling, but dependent of each other in a way which may actually act as an obstacle to evolution.

5.2.2. Engineering

Engineering in clean technology is a matter of applying different technical skills to achieve environmentally beneficial effects without undue economic sacrifices. Most often, it is a combination of many techniques rather than the utilization of a single fundamental principle. In fact, the further we push technology toward the limits imposed by nature, the more complex the engineering tasks we must face. This is particularly true in process engineering, and oddly enough, as the problems to be solved grow more complex, the most efficient way to attack them seems to be to simplify the process, taking into account the new constraints offered by the environmental limitations in addition to the former economic ones.

It is safe to say that the larger the number of operations that a process consists of, the more losses it will entail. Assume, for example, a production consisting of seven consecutive steps (not unusual in a production chain). Performing each step at an efficiency of 80% gives a total efficiency of $0.8^7 = 0.21$; if the number of steps can be reduced to, say, four, the total efficiency will be 0.41, a marked improvement. To achieve the same effect without reducing the number of steps would require an efficiency of 0.88 for each step, a nearly impossible requirement in most cases.

An almost classical example of the efficiency of this type of nonwaste technology engineering, in which several effects are integrated into the same steps is the flash-melting process. When first introduced by the Finnish company Outokumpu Oy for copper ore refining some 40 years ago, this process represented something close to a revolution in the field and rapidly became a commercial success.[25] The driving force for the success at that time was increased emphasis on the economics of raw material handling, particularly reduction of electricity costs; the environmental benefits were an extra bonus. Today, the process has been expanded to the recovery of different metals, and the environmental benefits are no longer an insignificant side benefit, but sometimes the main reason for switching to this technology.

The sudden increase in the price of electricity due to relocation of production facilities in the turbulence of World War II occasioned the decision to develop a new copper-melting process. Evaluation of the existing possibilities led to the decision to develop a process in which the fuel value of the sulfidic copper concentrate would be utilized to the maximum. Such so-called autogenous processes were known from the theory of metallurgy, but no practical applications were known at the time.

The principle of the flash-smelting process is very simple. The dried concentrate is burned in an air/oxygen suspension in such a way that the liberated oxidation energy can be utilized for the melting of the concentrate. Three separate, conventional processes are combined into a single flash-smelting unit: roasting, smelting, and partial converting.

The flash-smelting furnace consists of three compartments (Figure 11). A rectangular lower furnace equipped with cylindrical shafts at both ends. Sand and concentrate are fed into the first cylindrical feed shaft, in which an optimal suspension is formed together with the combustion air. The concentrate burns and partly melts in this feed or reaction furnace. The final melting occurs in the lower furnace in which the slag and (metal) matte also separate. The vital control of the process is done by controlling the temperature or adjusting the air/oxygen ratio.

The off-gases from the furnace are taken through the second shaft furnace to a waste-heat boiler. The particulate material not trapped in the furnace are separated by electrostatic precipitation and returned to the process. The sulfur dioxide-rich gas is conducted to the sulfuric acid and sulfur dioxide concentration unit. In the future, sulfur emission control will become tighter all over the world. Concerning the copper smelting, Outokumpu is prepared to meet the new challenges. Upgrading from Pierce-Smith converting to the Kennecott-Outokumpu flash converting means technical and economic advantages similar to those achieved by the replacement of a reverberatory furnace by a flash-

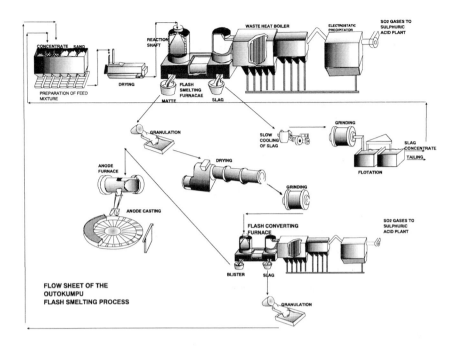

Figure 11.   Flowchart of the Harjavalta copper smelter. (Courtesy of Outokumpu Oy.)

smelting furnace. Flash converting, like flash smelting, offers an unusual combination of benefits: better sulfur recovery and reduced capital and operating costs. The efficiency of this early example of nonwaste technology has been widely recognized, and today, one third of the world's copper is produced by this method. The principle has further been extended to both nickel and lead smelting.

Perhaps this pioneering example of new engineering trends is a fitting introduction, as metallurgy is one of the oldest industrial activities of man and is still the backbone of our industrial society. There has been forecasts of a gloomy future for metallurgy and metallic materials. New synthetic materials have indeed developed and are replacing many of the traditional uses of metals, but this has happened mainly in areas where the properties of metals have not been utilized fully. From time to time, the shortage of mineral raw materials has also been an element of concern.

Despite these menaces, the growth of metals production and the development of metallurgic processes have been rather spectacular over the past decades. In addition to the increase in production volume (Figure 12), environmental concerns and energy savings have been a major concern in process development. Part of the development work has also been necessary in order to adjust the production to new raw materials. The present supply, although limited, seems sufficient on a moderate time scale. A very important raw material source, which is still growing, is the use of recycling. According to estimates, metallic scrap can meet about 30 to 50% of the raw material need, resulting in an enormous savings of energy.[26]

Blast furnaces have been used for iron making for roughly the last 500 years and are still the main process for iron making today, although the last decades have brought about important developments in the basic process. Coke consumption has decreased from about 1000 to about 400 kg/t of hot metal, and a value as low as 200 kg/t of hot metal, which is close to the theoretical limit, seems feasible. Today, the main line of development is toward the substitution of coal for the more expensive coke.

The basic principle in this development is first to make the prereduction using the reducing off-gases from the final reduction smelting reactor, which resembles the lower part of a traditional blast furnace. A scheme of the typical process route is presented in Figure 13.

Several development programs are underway, and at least one industrial-scale unit is in operation in South Africa (using the Austrian-German COREX process).

Environmental concerns and waste reduction are also the driving forces for the process development of metals production (Cu, Ni, Zn, Pb, Co, etc.) from sulfidic raw materials. The line of development, apart from hydrometallurgical routes, which are mainly developed for complex and low-grade ore materials, is the so-called "intensive smelting processes", of which the Outokumpu flash-smelting process was an early example.

A remarkable example of how new technical development can penetrate rather quickly even in a capital-intensive, commodity-producing domain such as steel making is provided by the rapid introduction of continuous casting, which today accounts for nearly 100% of the market (Figure 14).

Further development of the process is underway. For steel, the first stage is thin-slab casting, which produces 30 to 50-mm-thick slabs,

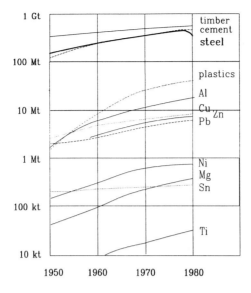

Figure 12. Production of steel, some metals, and nonmetallic materials.[26] With permission. (From Duckworth, W. F., The challenge to the materials, *Technol. Design,* 4, 924, 1984.)

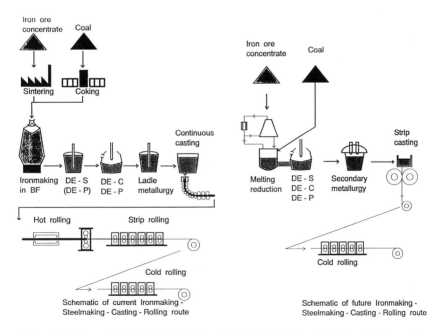

Figure 13. Current and future iron-making route. (From Holappa, L. E. K. and Jalkanen, H., NEUT/SUT-HUT Symp. Process Metallurgy, China, May 23–25, 1990. With permission.)

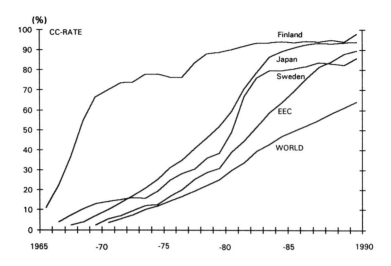

Figure 14. Continuous casting rate in a few countries and the European Community (EEC countries). (From Holappa, L. E. K. and Jalkanen, H., NEUT/SUT-HUT Symp. Process Metallurgy, China, May 23–25, 1990. With permission.)

which decreases the need for rolling considerably (Figure 15). A more radical development is strip casting, in which a continuous strip of 20 to 1 mm (or even less than 1 mm) is produced. The thinner the dimension, the more rapid the solidification. Strip casting results in a drastic simplification of the whole process (Figure 15) and also makes smaller production units economically feasible.

Interesting development possibilities are also offered by the so-called powder casting or spray casting methods, in which the metals stream is first atomized and the semisolid particles then deposited on a substrate. Further off in the future are applications of magnetohydrodynamics and electromagnetic plasma techniques.

Although the general engineering trends brought about improvements before the needs of clean production techniques were even realized, all the possibilities have by no means been exhausted. In fact, the combined effects of sophisticated technology and environmental awareness are powerful allies today.

The fact that reuse of metal scrap is a way of saving both nonrenewable resources and energy has long been recognized in the industry concerned. The energy advantage is especially pertinent in the case of aluminum, for which the production of virgin material requires large amounts of

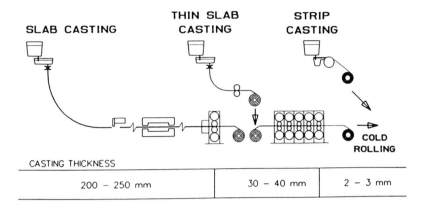

Figure 15. Comparison of new casting methods with conventional slab casting and rolling. (From Holappa, L. E. K. and Jalkanen, H., NEUT/SUT-HUT Symp. Process Metallurgy, China, May 23–25, 1990. With permission.)

(cheap) electricity. This has also been well recognized, and consequently the recirculation rate of aluminum metal is already well established in many countries.

Recently, a new method, reducing even further the energy needed for processing aluminum scrap, was introduced by an Italian company. The process involves direct extrusion of the aluminum scrap, requiring no resmelting. After being appropriately sized and dried, the scrap is directly extruded by means of a continuous technology.[27]

During the various stages of aluminum production from liquid metal to final products, a lot of trimmings are generated from machining operations.

Recycling these trimmings causes some problems because of their small size and the presence of oil and moisture deriving from cooling emulsions utilized in the machining operations. Moreover, trimmings from different aluminum alloys are often mixed together.

A group that produces and transforms 250,000 t/y of primary aluminium in its plants has a generation of about 6000 t/y of trimmings. Using a conventional recovery process (Figure 16) 1000 to 1500 t/y of metal are lost corresponding to lost revenues of 1.5 million U.S. dollars annually. Within the new process, recycling of trimmings is carried out without any melting phase (Figure 17).

After previous preparation steps, trimmings are directly and continuously converted into extruder products by a conform machine.

**RECYCLING OF ALUMINIUM SCRAPS WITH CONVERSION
INTO EXTRUDED PRODUCTS BY CONVENTIONAL SYSTEM**

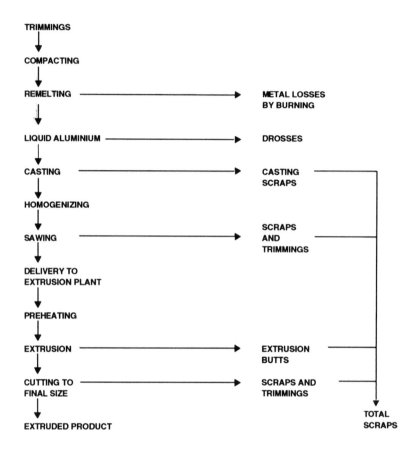

Figure 16. Flow chart of conventional process for recycling trimmings. (From Lazzaro, G. and Atzori, C., Technical Report from ALURES S. C. p. A. — Centro Technico Processi, Portoscuso (CA), Italy, With permission.)

The new recycling process is therefore characterized by fewer steps, higher recovery efficiency, and low generation of new scraps (Figure 18). Processing 100 kg of trimmings using the conventional method produces about 55 kg of extruded product, whereas with the new method about 95 kg are produced. The experience from a two year test period indicates that about 0.6 U.S. dollars per kg are saved using the new process as compared with the conventional method.[38] The new process, which won a clean technology award in Italy in 1989, consumes

RECYCLING OF ALUMINIUM SCRAPS WITH DIRECT AND
CONTINUOUS CONVERSION INTO EXTRUDED PRODUCTS

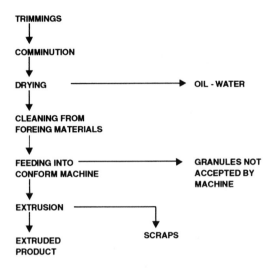

Figure 17. Flow chart of the new process for direct and continuous conversion of trimmings into extended products. (From Lazzaro, G. and Atzori, C., Technical Report from ALURES S. C. p. A. — Centro Technico Processi, Portoscuso (CA), Italy, With permission.)

73% less energy than the traditional method of recycling aluminum scrap, because it eliminates three energy-intensive steps: resmelting, homogenization of the billets, and preheating of the billets prior to extrusion. In addition, practically no metal is lost in the process, whereas 20% is lost through oxidation in the traditional scrap metal cycle. The process, which combines savings with environmental benefits, reduces combustion fumes by 90% and eliminates slag and polluting residues altogether.

An example of the power of simplification offered by new technical breakthroughs in another well-established technological branch (often labeled by the, in this context, rather negative description "mature") is given by Gullichsen.[28]

In a conventional process, 1700 tons of water are handled for each ton of pulp produced only as a transportation fluid for the fibers from one unit operation to another (Figure 19). Obviously, a reduction of the circulatory water volumes is in line with the goals of clean engineering.

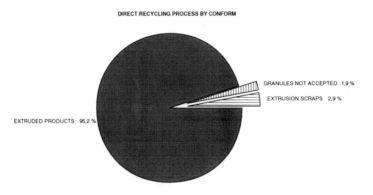

CONVENTIONAL RECYCLING PROCESS

CASTING SCRAPS 8 %
EXTRUSION BUTTS 2.2 %
DROSSES 10 %
EXTRUSION SCRAPS 15.6 %
MELTING LOSSES 15.6 %
PRODUCTS 54.2 %

DIRECT RECYCLING PROCESS BY CONFORM

GRANULES NOT ACCEPTED 1,9 %
EXTRUSION SCRAPS 2,9 %
EXTRUDED PRODUCTS 95,2 %

Figure 18. Comparison between material flows in a conventional alumina recovery process and the conform process. (From Lazzaro, G. and Atzori, C., Technical Report from ALURES S. C. p. A. — Centro Technico Processi, Portoscuso (CA), Italy, With permission.)

Remarkable progress in this respect has been achieved through the introduction of medium-consistency technology (Figure 20). The development of high-consistency pumping of pulp suspension offers the potential for total elimination of this water demand of a pulp factory (Figure 21). It is easy to imagine all the advantages in process and energy economy of such a simplification.

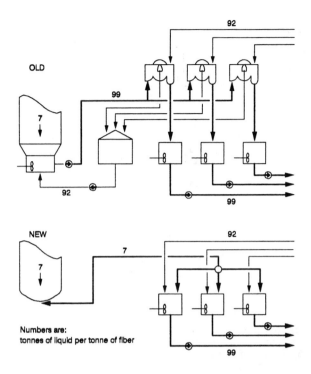

Figure 19. Conventional bleached kraft fiber line. Batch cooking, drum washing and drum washer bleaching.[28]

5.3.   PHOTOCHEMISTRY

In the area of clean process engineering, two domains of interest surpass the others:

1. Reduction of waste generation
2. Increased energy efficiency, preferably through the use of renewable energy sources

A totally solar-driven industry constitutes, in many respects, the ultimate goal of sustainable development. Few persons reflect on the fact that, in a broader sense, we are already there. More than 90% of the world's primary energy demand already is a result of solar-driven

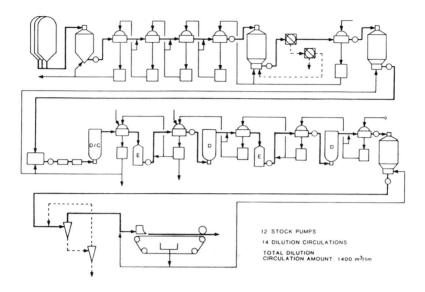

Figure 20. Demonstration of simplifications reached in paper mill pulp handling when converting to medium-consistency technology. (From Gullichsen, J., *VTT Symposium, 102,* 1989, 259. With permission.)

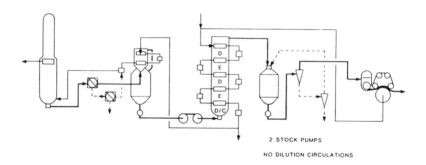

Figure 21. Potential future fiber line. (From Gullichsen, J., *VTT Symposium, 102,* 1989, 259. With permission.)

reactions. Biomass, coal, mineral oil, and natural gas are all storage forms of solar energy (in a longer time perspective, the remaining 10%, consisting of nuclear and geothermal activity, can also be considered a legacy of solar origin).

The principal ways by which solar chemistry can enter into modern society in a growing manner are:

- Direct utilization of solar thermal energy
- Conversion of solar radiation to electricity
- Utilization of solar radiation for chemical reaction

It has been proposed that the photon flux in solar radiation can be used directly to drive quantum processes in chemistry[29] and achieve reaction paths that can be difficult to obtain otherwise. The sun can be considered a black-body emitter of 5777 K, for which, consequently, in the short wavelength of the spectrum, the chemical potential of the photon flux is high.

The chemical potential μ (ν) is a measure of the driving force of absorbed radiation in a photochemical reaction.[30]

$$E_\nu(\nu) = \frac{2\Omega * h\nu^3}{c^2} \frac{1}{\exp\left(\dfrac{h\nu - \mu}{kT}\right) - 1}$$

E_ν (ν) is the spectral flux density of the solar radiation, $\Omega*$ is the solid angle, and T is the temperature of the system.

The chemical potential is, by rearrangement:

$$\mu(\nu) = h\nu - kT \cdot l_u\left(1 + \frac{2\Omega * h\nu^2}{c^2 E_\nu(\nu)}\right)$$

where $h\nu$ is the photon energy at frequency ν or wavelength $\lambda = c/\nu$. By optical concentration, this high chemical potential can be extended to cover the entire spectrum (Figure 22).

The direct use of solar fluxes requires absorption by the chemical process; generally, this absorption, determined by the molecular (electronic) structure of the molecule, is limited to a very narrow and specific band of the spectrum, which limits the energy efficiency of the process.

5.4.   THERMOCHEMISTRY

Thermal reactions differ from photochemical reactions in that there are no limiting selection rules for temperature changes[31] (Figure 23).

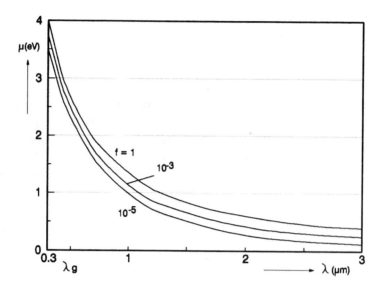

Figure 22. Chemical potential of the solar flux f denotes the dilution of the radiation (ratio of flux density to maximum flux density at surface of sun).

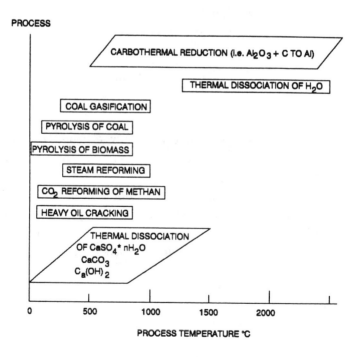

Figure 23. Temperature ranges of industrial thermochemical processes.

When employing solar energy for thermochemical reactions, the procedure is identical to that employed for thermal solar energy generation; the radiation is concentrated and then absorbed in receivers. The temperature attained can be as high as 3000°C, and theoretical heat fluxes of up to 63 MW/m^2 can be reached.[32]

Solar-driven thermochemistry is essentially conventional thermochemistry, as the origin of the heat is irrelevant for the reaction itself; the only important thing is to reach (and maintain) the required temperature. The particularities in terms of engineering relate to the transient nature of the solar heat source, i.e., variations due to atmospheric conditions as well as seasonal variations, which may call for special technical solutions.

Interesting and in some cases unique reaction conditions can be created by the simultaneous use of the thermal excitation and photochemical activity offered by the nearly 6000-K solar emitter. Thus far, this area of development has been explored very little, but thermodynamic arguments[31] indicate that advantages can be achieved, such as lowering of the reaction temperature of endothermic reactions (Figure 24).

A reaction of profound interest for solar-driven chemistry is the decomposition of water to its basic constituents, hydrogen and water. If technically feasible such a process would open up the ideal route to the clean-energy carrier, hydrogen, which could be rather easily adapted to the present energy society. In addition to being a potential energy carrier, hydrogen also is of interest as a chemical reactant (this is also true for the oxygen simultaneously produced).

As demonstrated earlier, the thermochemical decomposition of water would require very high temperatures (on the order of 4000°C), involving difficulties concerning materials as well as engineering. Thermochemical multistep processes have been proposed to overcome this difficulty, and to perform the water splitting at a lower temperature.[34] An unavoidable consequence of this temperature reduction through multistage reactions is that a larger portion of the heat introduced into the reaction must be sacrificed as waste heat carried away by the secondary reactions(s) (Figure 25).

5.5. ENERGY SAVING

Apart from the guidance of technology toward the use of more environmentally benign processes and materials, the efficiency in the

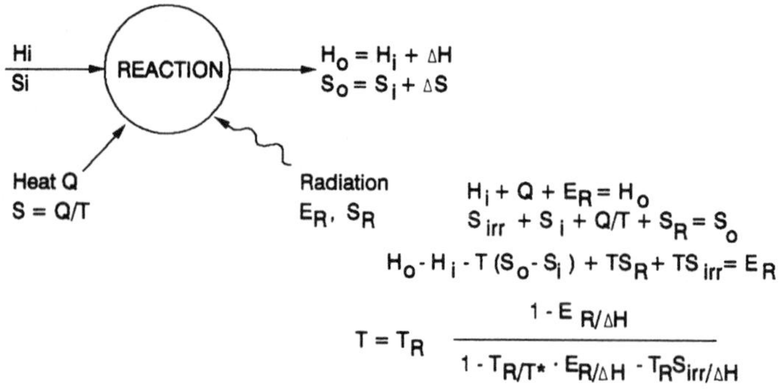

Figure 24. Lowering of reaction temperature by combining a thermal reaction with an intense photon flux.

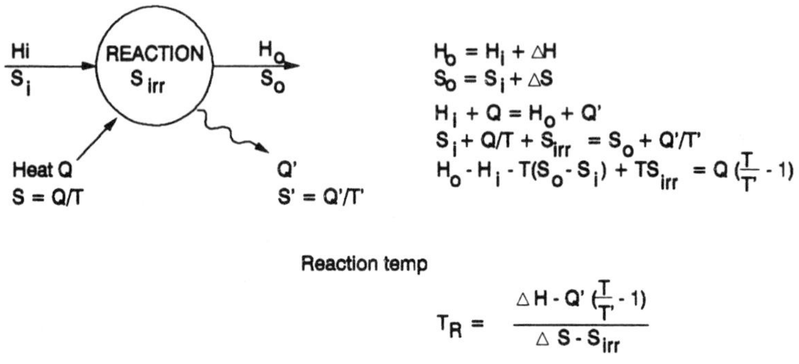

The equations associated with Figure 24:

$$H_o = H_i + \Delta H$$
$$S_o = S_i + \Delta S$$

$$H_i + Q + E_R = H_o$$
$$S_{irr} + S_i + Q/T + S_R = S_o$$
$$H_o - H_i - T(S_o - S_i) + TS_R + TS_{irr} = E_R$$

$$T = T_R \frac{1 - E_{R/\Delta H}}{1 - T_R/T* \cdot E_{R/\Delta H} - T_R S_{irr}/\Delta H}$$

The equations associated with Figure 25:

$$H_o = H_i + \Delta H$$
$$S_o = S_i + \Delta S$$
$$H_i + Q = H_o + Q'$$
$$S_i + Q/T + S_{irr} = S_o + Q'/T'$$
$$H_o - H_i - T(S_o - S_i) + TS_{irr} = Q\left(\frac{T}{T'} - 1\right)$$

Reaction temp

$$T_R = \frac{\Delta H - Q'\left(\frac{T}{T'} - 1\right)}{\Delta S - S_{irr}}$$

Figure 25. Two-stage reaction.

use of raw materials, especially those related to primary energy production, are the essence of clean technology. The two dominating principles are:

1. To use as little as possible of the virgin raw material for a given product or service

2. To direct as far as possible the inevitable residues or waste streams obtained in the processing to useful purposes

These concepts are especially pertinent as regards energy use, a domain where perhaps this type of thinking has penetrated deepest, but they are also valid in a more general sense for all raw materials.

As has often been stated from an environmental point of view, the purest form of energy is one which is not used at all. Using as little energy as possible for a given service is obviously the best form of environmental technology. It is perhaps worth pointing out that this is not the same as minimizing total energy use, as many purification processes require energy.

Energy efficiency and the use of waste heat is particularly well developed in industry, where the streams are large and the technical solutions can easily be justified by increased revenues for the companies. The potential of large-scale integration of this kind has been well demonstrated by the increased efficiency in electricity production through the combined cycle power generation process (see Chapter 4).

Another technology which offers opportunities for greatly enhanced fuel efficiencies is cogeneration, where thermal electricity production is combined with the use of waste heat for domestic heating, or other purposes where low-temperature steam (or in some cases, warm water) can be used. While a typical fuel efficiency for electricity production only (using conventional technologies) falls in the range of 20 to 35%, the fuel efficiency for a cogeneration system can exceed 80% (Figure 26).

This type of integration does, however, assume that there is a balanced demand for both heat and electricity, which is not always the case; in fact, energy saving (reducing heat losses through increased insulation and other technical improvements) continuously distorts the demand, putting more pressure on electricity generation, the most valuable form of energy (and hence the need to find ways to increase electricity production or to invent other means of utilizing the low-grade energy of lower temperature streams, i.e., heat pumps).

As mentioned earlier, the industrial sector is rather well developed in this respect, as the incentive for improved efficiency is clear. This is not so in the consumer sector, whose market share of electricity use is increasing in most industrialized countries. The economic incentive is there, of course, through the electricity bills, but as the price of electricity is still very low compared to many other things in our affluent society, it is not a very effective instrument.

Moreover, consumer groups, although composed of individual small users, may occasionally show collective and not always predictable behavior. They are also less inclined to accept limitations than the

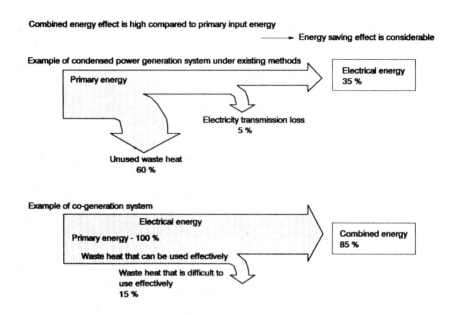

Combined energy effect is high compared to primary input energy

—————→ Energy saving effect is considerable

Example of condensed power generation system under existing methods

Primary energy

Electricity transmission loss
5 %

Electrical energy
35 %

Unused waste heat
60 %

Example of co-generation system

Electrical energy

Primary energy - 100 %

Waste heat that can be used effectively

Waste heat that is difficult to
use effectively
15 %

Combined energy
85 %

- This efficiency does not include domestic energy consumption and energy consumption for environmental measures
- This efficiency is ideal when there is an optimum heat demand-power demand combination
- The efficiency of electric power systems can improve considerably if air-conditioning uses an electric heat pump

Figure 26.   Energy saving effect of cogeneration.

industrial sector, as liberation from constraints is one of the underlying reasons for the growing energy demand in the consumer sector. This behavior puts additional strain on the power generation machinery, which is designed to operate most economically on a given base-load. Increased capacity, the so-called peak production, is much more expensive. The high cost of peak power production is the reason why improvement of the energy efficiency of household appliances is important despite the fact that they are rather modest electricity users. This is also one of the reasons why small-scale power generation (decentralized system) becomes increasingly attractive.

The combined effect of multiple users can sometimes be quite impressive. For example, if all illumination in Finland were converted from ordinary incandescent lightbulbs to glow discharge lamps, 1.5 to 2 TWh of electricity would be saved, corresponding to some 2.5 to 3% of the total, or roughly one large power plant — without a marked differ-

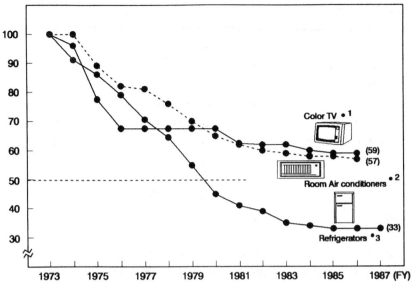

Unit: Electric power consumption in 1973 = 100

*1 indices of electric power consumption of the 19,20 inch type
*2 indices of electric power consumption of the separate type
*3 indices of electric power consumption for one month
   (annual average) of the 2-door, 170 liter type

Figure 27. Progress of energy conservation in major home appliances (From *Energy Conservation in Japan,* JETRO, 1990. With permission.)

ence in the service provided to the consumer, one might add. In fact, quite impressive technical achievements in reducing the electricity consumption of household appliances have been realized over the last decade without the individual consumer being affected or perhaps even noticing the difference (Figure 27).[35]

Heat pumps offer another opportunity to upgrade low-grade waste energy for useful purposes. Although the principle has been known for a long time, modern technology or correct pricing of energy may increase the prospects for small-scale heat pumping in the domestic sector.

5.6.   ENERGY STORAGE

It was mentioned earlier in the context of cogeneration that a prerequisite for such a system was a balanced demand for electricity and heat, and that this condition was not always met. The same is true in a more general sense when speaking of renewable energy production. These resources are often seasonal and most abundant during periods when energy is not necessarily needed.

A central scientific and technological task is to find ways of storing this often abundant, but dispersed energy in a concentrated form from which it can be liberated at will. As mentioned earlier, fossil fuels are solar energy that has been stored for many millions of years; even the nuclear fuels uranium and, if fusion becomes reality, deuterium and tritium are legacies from solar activities (big bang) way back in cosmic times. For the production of these, human technologies are powerless; however, oil and other liquid fuels (and, of course, charcoal) can already be produced from their natural precursor biomass, in essence by mimicking the reactions that resulted in the formation of the fossil reserves. Along these lines, enhanced production of biomass is still an area of technical interest for storing solar energy.

An illustration of the crucial importance of efficient energy storage is provided by the following example.[36] If the earth's population is assumed to stabilize at a level of 10 billion and we would like to provide all persons with an adequate power supply of, say, 4 kW each, a surface area corresponding to only 18% of the Sahara Desert would suffice provided we could trap and store only 10% of the incident solar radiation. The comparative volume requirements for different energy storage systems are shown in Table 8.

Storing heat in large heat reservoirs is a direct form of energy storage currently under development. The solar pond is a particularly simple device for low-temperature heat collection and storage. It consists of a shallow pond with a salt concentration that increases with depth. Solar radiation is absorbed by the pond, and heat losses are minimized by the absence of convection currents. In Israel, such ponds reach about 85°C, and a heat exchanger placed at the bottom can provide heat to a Rankin engine that drives a generator.[37] The thermal efficiency is only 8%, and as the optical collection efficiency is only 20%, the total efficiency is low. Nevertheless, the solution might be viable in a broader context, taking into account its simplicity and the economy of land use.

Table 8
Volume Requirement of 10^4 kWh of Energy Stored in Different Systems

Chemical energy	1 m³ oil
	3 m³ NH₃
Electrochemical	600 m³ lead/acid
	40 m³ Zn/air
Latent heat	10 m³ NaF/CaF₂/MgF₂ at 800°C
Sensible heat	250 m³ of warm water at 90°C used at 60°C
Mechanical	Fly wheel: 3 tons spinning at 10^4 t/min (r = 5 m);
	hydro-electric: 3600 m³ of water at Δh = 100 m

Energy Density of Some Fuels

Substance	kWh/kg	kWh/m³
Hydrogen	34–40	3.3 gas
		2300 liquid
Methane	18–20	10–11 gas
Ethanol	7.5–8.3	6×10^3–6.6×10^3
Gasoline	11.7–12.8	8.4×10^3
Mineral coal	7.6–9.7	20,000
Wood	4–5	5,000

Note: All our fuels consist of stored chemical energy. They are burned with oxygen from the air. It would be far more economical to use them in fuel cells.

REFERENCES

1. Hanny, J. B. and Hogarth, J. On the solubility of solids in gases, *Proc. R. Soc. London*, 29, 324, 1879.
2. Gährs, H. J. High pressure extraction — State of the art, present applications and future prospects, in *Separation Technology for Fine Chemicals IV*, A. Rapport, 1988, 354.
3. Huggin J. Membrane technology has achieved success, yet lags potential, *Chem. Eng.*, October, 22, 1990.
4. Ben Aim, R. and Vladan, M. La place des techniques à membranes dans les procédés propres, *UNEP ind. Environ.*, 12(1), 15, 1989.
5. Saari, M. *Prosessiteollisuuden Erotusmenetelmät*, VTT Res. Note (and references therein), Technical Research Centre of Finland, 1987, 730.
6. Marr, R., Lackner, H., and Draxler, J. Industrial application of liquid emulsion technique, in *VTT Symposium 102*, Vol I, 1989, 345.
7. Anderson, R. A., and Sherman, J. D. Molecular sieve adsorbents and ion exchanger historical review, recent progress and future directions, *AIChE Symp. Ser.*, 80, 118, 1984.
8. Remy, H. *Treatise on Inorganic Chemistry*, Elsevier, New York, 1956.
9. Dwyer, J. and Dyer, A. Zeolites for industry, *Chem. Ind.*, April, 237, 1984.
10. Milton, R. M. U.S. Patent 2,882,433, 1959.

11. Ruthven, D. M. Zeolites as selective adsorbants, *Chem. Eng. Progr.*, February, 22, 1988.
12. Sweed, N. H. Parametric pumping and cycling zone adsorption — a critical analysis, *AIChE Symp. Ser.*, 80, 44, 1984.
13. Goto, S. and Matsubara, M. Extraction parametric pumping with reversible reaction, *Ind. Eng. Chem. Fundam.*, 16, 193, 1977.
14. Grevillot, G. ym. Thermofractionation: a new class of separation processes which combine the concepts of distillation, chromatography and the heat pump, *Int. Chem. Eng.*, 22, 440, 1982.
15. Rachez, D. ym. Stagewise liquid-liquid extraction parametric pumping. Equilibrium analysis and experiments, *Sep. Sci. Technol.*, 17, 589, 1982.
16. Wankat, P. C. Liquid-liquid extraction parametric pumping, *Ind. Eng. Chem. Fundam.*, 12, 372, 1973.
17. Chen, H. T. ym. Separation of proteins via semicontinuous pH parametric pumping, *AIChE J.*, 23, 695, 1977.
18. Chen, H. T. ym. Separation of proteins via multicolumn pH parametric pumping, *AIChE J.*, 26, 839, 1980.
19. Chen, H. T. ym. Parametric pumping with pH and ionic strength: enzyme purification, *Ind. Eng. Chem. Fundam.*, 20, 171, 1981.
20. Chen, H. T. ym. Semicontinuous pH parametric pumping: process characteristics and protein separations, *Sep. Sci. Technol.*, 16, 43, 1981.
21. Rice, R. G. and Foo, S. C. Continuous desalination using cyclic mass transfer on bifunctional resins, *Ind. Eng. Chem. Fundam.*, 20, 150, 1981.
22. Rice, R. G. Progress in parametric pumping, *Sep. Purif. Methods*, 5, 139, 1976.
23. Wankat, P. C. Cyclic separation techniques. Teoksessa: percolation theory and applications, Alphen aan den Rijn Sijthoff and Noorhoff, 1981, 443.
24. Kuyucak, N. and Volensky, B. New biosorbents for non-waste technology, in *VTT Symposium 102*, Vol. 1, 1989, 313.
25. Sulanto, J. Outokumpu flash melting — an efficient and non-polluting metallurgical process, in *VTT Symposium 103*, Vol 2, 1989, 55.
26. Holappa, L. E. K. and Jalkanen, H. Outline of research and development of metallurgical processes today and in the near future, in NEUT/SUT-HUT Symp. Process Metallurgy, China, May 23 to 25, 1990.
27. Lazzaro, G. and Atzori, C. Recycling of Aluminum Trimmings by Conform Process, Technical Report from ALURES S.C. p.A. — Centro Tecnico Processi, Portoscuso (CA), Italy.
28. Gullichsen, J. Utilization of resources of the wood pulping industry — a holistic view, in *VTT Symposium 102*, 1989, 259.
29. Calzaferri, G. Photochemical, electrochemical and thermochemical transformation and storage of solar energy: thermodynamic aspects, in *Solar Energy '85: Resources — Technologies — Economics*, ESA-SP-240, proc. summer school held at Igls, Austria, July 31 to August 9, 1985, European Space Agency, 1985, 93.
30. Sizmann, R. Solar radiative energy, in *Solar Energy '85: Resources — Technologies — Economics*, ESA-SP-240, proc. summer school held at Igls, Austria, July 31 to August 9, 1985, European Space Agency, 1985, 5; Minder, R., Wolf, M., and Leidner, J. R. *Chimia*, 42, 124, 1988.

31. Nix, G. and Sizmann, R. High temperature, high flux density solar chemistry in Proc. Santa Fe Conf. High Temperature Technologies, Golden, CO, 1988, 40.

32. Hubbard, H. M. Scientific and technical challenges in electricity generation, *Chimia,* 43(7-8), 197, 1989.

33. Nakamura, T. Hydrogen production from water utilizing solar heat at high temperatures, *Sol. Energy,* 19, 467, 1977.

34. Sizmann, R. Solar driven chemistry, *Chimia,* 43, 202, 1989.

35. Europe-Japan Global Environmental Technology Seminar, 1990, 127.

36. Calzaferri, G. in Solar Energy '85: *Resources — Technologies — Economics,* ESA-SP-240, proc. summer school held at Igls, Austria, July 31 to August 9, 1985, European Space Agency, 1985.

37. Spiers, D. *Solar Electricity and Solar Fuels,* Advanced Energy Systems National Research Program, NEMO Rep. 7; Tabor, H., *Sol. Energy,* 27, 181, 1981.

38. Lazzaro, G. and Atzori, C. Recycling of Aluminum Trimmings by Conform Process, Technical Report from ALURES S.C. p.A. — Centro Tecnico Processi, Portoscuso (CA), Italy.

CHAPTER 6

Waste

Human society (as described in Section 2) can thermodynamically be considered an open system. It extracts energy from the exterior and rejects waste into its surroundings, as does the ecosystem on the whole. There are, however, some important differences between modern industrial society and the ecosystem that are worth emphasizing:

1. The human mind has been able to conceive and construct materials and products foreign to the ecosystem and, as such, indigestible by it.

2. Perhaps most important, the scale of human operations has assumed proportions that far surpass those found in nature; in fact, many areas in industrialized countries have reached a state where the industrial activities (as evidenced, by energy densities) far surpass the natural activities.

The acute problem of waste management is a manifestation of both these effects. This concern is, of course, not new, although it has been largely neglected until recently, mainly due to insufficient knowledge about the intricate mechanisms between industrial society and nature, but also due to the rapid growth of economic activity in the last decades. We are speaking of large quantities of material due not only to industrial operations, but to all sectors of human activity (Tables 1 to 3).

6.1. INDUSTRIAL WASTE

The negative effects of a production site on its environment are, in the traditional sense, considered to be restricted to the effluents from the process. These are generally classified as emissions when occurring as rather dilute, contaminated streams of air or water, or as waste when we speak about more concentrated streams of liquid or solid residual matter from the process.

An exact definition of waste is lacking, and the term generally is employed in a rather loose fashion to signify a material stream without value to the process concerned. This does not, however, imply that the material would be totally worthless as such. On the contrary, many industrial waste streams are, or could be, valuable raw materials for other processes. The problem is the rate of production, transport, and very often lack of incentive due to negative attitudes. The purely technical reasons, such as impurities or the form of the waste, are often less significant and could generally be overcome. In an analysis of the major obstacles to waste minimization, Palmer[1] found that 60% were political in origin, 30% financial, and only 10% of a technical nature.

With the increasing public concern about the inadequacies of current waste handling, the attitudinal and political barriers are decreasing and the relative importance of the technical factors have increased. This mounting concern among the citizens is reflected in a recent poll taken within the EEC countries which shows that the European citizen of 1985 is more concerned about his environment than he was in 1982 (Table 4).

Much current research and development activity is devoted to efforts to reduce and reuse industrial (and urban) waste streams. The figures do not reflect the fact that much of the industrial waste is, in effect, already recycled or reused. In a sense, this makes the residual problem much harder to solve, as most of the obvious or easy solutions already have been realized. The rising cost of waste handling, however, makes a further decrease of industrial waste an economic necessity for most

Table 1
Estimated Total Annual Wastes in the EEC in 1982[2]

Total waste including	2500
Agricultural waste	950
Sewage sludge	300
Industrial waste	180
Household waste	150

Note: Waste in millions of tons.

Table 2
Estimated Quantities of Waste Produced by Industry[2]

Country	Industrial waste	Hazardous waste
Belgium	8	1.5[4]
Flanders	5	1
Denmark	0.8	0.1
Germany	52	4.5[5]
France	38	2[6]
Ireland	1.2	0.075
Italy	35	2–5
Netherlands	5.1	1
U. K.	30–40	5

Note: Waste in millions of tonnes per year.

Table 3
Estimated Quantities of Waste Produced by Chemical Industry[2]

Country	Industrial waste[7]	Hazardous waste[a]	
Belgium	2.8	1.0	0.9
Flanders	2.4	0.9	0.8
Denmark	n/a	0.04	0.12
Germany	10	3.0[8]	(3.0)
France	10	0.5–0.8[8]	1.8
Ireland	n/a	(0.04)	0.08
Italy	3	0.9	1.3
Netherlands	1.1	(0.5)	0.8
U. K.	5–10	2–5	1.6

Note: Waste in millions of tonnes per year.

[a] From different sources and estimates.

companies. In industrial production, virtually the only remaining option is to decrease the formation of waste streams at the source by internal process streams. This may involve a change of raw materials as well as,

Table 4
Attitudes Within the EEC Countries Toward Environmental Questions: 1985 and 1982

	Very	Fairly	Some-what	Not at all	Don't know	Total	Index for EURIO 1985	1982
The way in which industrial waste is disposed	47	32	12	6	3	100	2.23	2.18
Damage to marine organisms and beaches	45	37	11	5	2	100	2.23	2.21
Pollution of water, rivers, and lakes	43	38	12	6	1	100	2.20	2.02
The extinction world-wide of plant or animal species	42	37	14	5	2	100	2.19	2.01
Air pollution	41	36	14	7	2	100	2.13	1.96
Possible changes in the world's climate owing to carbon dioxide produced by burning coal and oil products	38	33	16	8	5	100	2.06	1.86
Exhaustion of world's natural resources	35	37	18	7	3	100	2.04	2.02

Note: Values cited represent level of agreement (in percentile) with statement given.

in the end, redesigning the product itself. The potential of such new engineering for improving the environment is considerable, as demonstrated by several successful examples. Virtually the only limitation is that posed by our lack of imagination. One might even go so far as to state that here lies the seed for technical rejuvenation of the industrial establishment that is necessary for continued prosperity on a sustainable basis.

6.2.   HAZARDOUS WASTE

The formation of toxic or hazardous waste constitutes a problem of its own. Although the exact definition of this class of waste varies from country to country, the name itself is self-explanatory: we are dealing with wastes that present a real danger to human health or the environment.

As a result of neglect until some 15 years ago, improper deposition and abandoned waste-dump sites containing hazardous wastes present a major pollution problem in many countries today. Because of these problems, most industrialized countries have adopted laws regulating hazardous waste management. The first European Economic Community Directive on Toxic and Dangerous Waste was implemented in 1980. The Organization for Economic Cooperation and Development (OECD) adopted its first Decision on the Export of Hazardous Waste from OECD countries in 1986. The Cairo Guidelines and Principles for the Environmentally Sound Management of Hazardous Wastes were adopted by the United Nations Environment Program (UNEP) in 1987. Many countries, however, still lack the legal framework for dealing properly with hazardous wastes.

Although the regulations differ from country to country, there are certain basic similarities. In the regulations, the scope of application is indicated by giving general information about the waste or by listing separately the wastes considered to be hazardous, using criteria such as origin of the waste, toxic substances present, and their properties. Up to now, no generally accepted list of toxic substances has been agreed upon, although such work is in preparation.

Many countries rely on the principle that the generator of the waste is responsible. Generally, this results in various procedures to ensure a cradle-to-grave management of the hazardous waste, i.e., an extended control from its generation to its proper disposal. Such legislation normally specifies that the waste generators must ensure that the waste produced is properly transported and disposed of, even if these tasks are subcontracted to someone else.

The Directives of the Economic Community of Europe cover such matters as intracommunity shipment and export of hazardous wastes from the Community[3]. These directives provide a framework for the proper handling of hazardous waste, but must be amended to follow developments in the field. In particular, work is in progress to amend the directive on waste and that governing the disposal of toxic and dangerous waste in relation to the definition and identification of waste, and to the promotion of clean technologies and "ecological" products.[4]

These provisions include the member states' obligation to designate competent authorities to be responsible for the supervision and administration of operations for the disposal of toxic and dangerous waste. These authorities, responsible in given areas, must plan, organize, and supervise such operations. They are responsible for issuing permits for the storage, treatment, and/or deposit of toxic and dangerous waste, and for controlling such undertakings and those responsible for the transport of the waste.

When waste is transported to a disposal center, it must be accompanied by an identification form.[5] Before any transfrontier shipment of hazardous waste takes place, the proper authorities must be notified and receipt of the notification must be acknowledged. However, in the case of identical and regular shipments of nonferrous metal waste intended for recycling, only a general notification is required. The notification must be accompanied by the following information:

1. The source and composition of the waste, including the producer's identity and, in the case of waste from various sources, a detailed inventory of the waste and, where such information exists, the identity of the original producers

2. Arrangements with respect to routes and insurance against damage to third parties

3. The measures to be taken to ensure safe transport and, in particular, compliance by the carrier with the requirements for such transport operations, as laid down by the member states concerned

4. The existence of a contractual agreement with the consignee of the waste, who should possess adequate technical capacity for disposal of the waste in question under conditions presenting no danger to human health or the environment[6]

REFERENCES

1. Palmer, P. *UNEP Ind. Environ.,* 12(1), 5, 1989.
2. Junger, J. M. and Lefevre, B. *UNEP Ind. Environ.,* 11(1), 15, 1988.
3. Notification must be made by means of a uniform consignment note, as established in Commission Directive 85/469/EEC of July 22, 1985.
4. Directive 78/319/EEC on toxic and dangerous waste establishes the general rules for the management of toxic and dangerous waste, which is defined as any substance or object which the holder disposes of pursuant to the provision of national law and containing or contaminated by one or more of the substances or materials listed in table and of such a nature as to constitute a risk to health or the environment.
5. Directive 84/631/EEC on the supervision and control of the transfrontier shipment of hazardous waste within the European Community covers the shipment of hazardous waste between member states, waste imported from or exported to non-Community countries, and waste in transit through the Community.
6. Where the waste is stored, treated, or deposited within a member state, the consignee must also possess a permit in accordance with Article 9 of Directive 78/319/EEC or Article 6 of Directive 78/403/EEC.

CHAPTER 7

System Analysis

7.1.   FLEXIBLE PROCESSES

Utilization of the economy of scale has to date been the predominant way of improving the economic efficiency of processes, particularly within the basic industry where the competition from (often) low labor cost or cheap raw material-producing countries is most strongly felt. This has often resulted in process units of gigantic proportions, causing severe disturbances in the social and ecologic surroundings. Also, such units require large capital investments, putting very high requirements on the efficiency of operation of the process, which results in pressure to operate the process under smaller and smaller profit margins to recover the investment as the competition tightens. Examples of such development can be found in many diverse industrial sectors such as petrochemistry, metallurgy, and pulping.

Due to the efficiency requirements, often linked to output volume, the large process units are rather inflexible and vulnerable to external changes such as changing market requirements or environmental legislation. On the other hand, the increasing pressure to save raw material requires the industry providing the basic raw material to further process this to an intermediate material which is as suitable for final production as possible. This leads to an obvious conflict between the requirement of maximal efficiency in terms of the output of the process and versatility in production, to provide a product which is as marketable as possible. This dilemma is clearly visible in heavy industries such as the metal foundry industry, where the trend is toward continuous foundry of close-to-final-shape profiles,[1] or paper manufacturing.[2]

The rapid pace of technological innovation and market restructuring favors smaller, more flexible manufacturing units that can operate at full capacity and lose little time in changing product quality, continuously providing higher-value specialized products. The powers of computer-integrated manufacturing (CIM) have clearly been manifested in flexible manufacturing systems, but the use of advanced automation and process control systems remains to be developed.

An example from the pulping industry, where these problems have been studied for some time,[2] indicates that the (at first sight) conflicting requirements of decreased downtime of the process when changing the product and increased customer-specific production can be overcome by applying sophisticated computer-integrated automation and process control. Although this example refers to a particular branch of industry, the general arguments are valid for many basic industries facing similar problems.

Depending on the product line, a pulp producer may have to switch between many different grades during the year in order to serve a variety of customers, resulting in less effective production time. Table 1 gives the cost structure-price flexibility of different pulp products calculated for a typical machine size of 200,000 t per year. From these, we can make the following rather general statement, having validity in a more general sense for industrial development. The advantage of fusing together several production lines (in this particular case, paper machines) under the same organizational strategy, is obvious. In this way, the company can obtain flexibility while still retaining the economy of scale on a single-machine level. This is done by a division of tasks between the machines, which allows for longer runs on a single unit. The savings potential corresponds to more than 100 Finnish Marks (about $25) in terms of production cost savings per ton (Table 2), and

Table 1
Cost Structure - Price Flexibility of Different Pulp Products Calculated for a 200,000 Ton/Year Machine Size

	Price[a]/ ton	Direct variable costs/ton	Capital costs/ton	Grade changes/ year	Time spent for grade changes (min)	% Total time
News, standard	2,000	1,500	300	30	450	0.01
News, multiproduct	2,500	1,600	330	400	12,000	2.5
LWC,[b] standard	2,800	1,900	350	40	800	0.02
LWC,[b] multiproduct	3,200	2,000	380	500	15,000	3
Fine, standard	3,500	2,300	400	250	7,500	1.5
Fine, multiproduct	4,500	2,500	450	700	28,000	5

[a] FIM, Finnish Marks.
[b] Light weight coating.

Table 2
Differentiation of Paper Products

	Value/ ton	Lost value added	Lost sales		% margins	Sensi- tivity for breaks	Break time %	Value of breaks	% of margins
			FIM	%					
N, S	189	189,000	378,000	0.1	1	Small	5	10 M	25
N, MP	5,050	5 M	12.6 M	2.5	12.6	High	8	40 M	40
LWC, S	340	410,000	850,000	0.15	1	Medium	6	34 M	34
LWC, MP	6,300	9.5 M	20 M	3	14	High	8	50 M	36
Fine, S	3,150	5.4 M	11 M	1.5	10	High	8	55 M	50
Fine, MP	11,800	23 M	53 M	6	17	High	10	90 M	30

Note: See Table 1 for explanation of abbreviations.

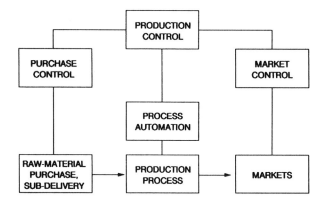

Figure 1. The interaction between functions included in the production control.

even more if we take into account the fact that longer runs also mean less downtime due to decreasing break sensitivity. This line of development is well under way, although not yet widely applied outside the paper-manufacturing industry. The most important milestones for process automation are presented in Table 3.

Due to improved data-linked networks, it no longer is necessary to place different production units in close physical proximity to each other. This opens up new dimensions of development that allow the advantages of scale to be combined with the benefits of decentralization mentioned earlier in this section.

Such flexibility for decentralization of production units without loss of efficiency opens up new avenues for technical and social development, with great impact on environmental questions as well. The described development requires automation solutions which give the users an integrated control of the complete process. This integration function is described in Figure 1.

7.2.   TRANSPORT SECTOR

The car is a decentralized, low-priced energy production unit. In industrialized countries, one car per two inhabitants is the equivalent of approximately 10 kW per person, which should be compared to a typical centralized power capacity of about 2 to 3 kW per person (order

Table 3
Implementation Strategies of CIM Technologies

Technology basis	Scope of control	Structure of system	Operative basis	Control room	Dominance period
Pneumatics and prepneumatics	Single loop, unit process	Decentralized, dedicated	Loop, sub-process control	Decentralized,	>1950s
Analog, electronic	Single loop, unit process	Decentralized, first-order functional integration	Process supervision	Central, process level	1960s
Analog/ digital (process computer)	Connected processes	Centralized, functional integration	Process operation and supervision	Central hierarchical	1960s–1970s
Digital/ digital (μp/process computer)	Plant	Decentralized, distributed functional integration	Plant-process operation	Central hierarchical	Late 1970s–1980s
26 digital (fully μp)	Integrated plant	Decentralized, distributed functional integration	Plant and process management	Central, hierarchical	Late 1980s–1990s
Future-information technology	Integrated: supply-plant distribution	Distributed fully horizontal and vertical integration	Plant management systems development, anticipating	Functional, integrated, horizontal distribution	1990s

Table 4
Energy Requirements for Passenger Transportation

Mode of transport	Maximum capacity (passengers)[a]	Time to travel 1000 km (h)	Passenger mileage (passenger-km/l)[b]	Energy Consumption (kcal/passenger-km)
Bicycle	1	6 d	660	12
Walking	1	30 d	200	37
Intercity bus	45	12	96	87
Subway train (10 cars)	1000	8	64	130
Local bus	35	—	45	186
Automobile[c]	4	10	20	404
Boeing 747	360	1	15	540
Concorde SST	110	0.5	6	1472
Ocean liner	2000	12	4	1940

[a] The relative effectiveness of various modes of transportation can be drastically altered if a smaller number of passengers are carried.
[b] Kilometers per liter of gasoline or the equivalent in food or other fuel: all values must be regarded as approximate.
[c] Typical American automobile, used partly for local travel and partly for long-distance driving.

Modified from Slesser, M., Energy in the Economy, Macmillian, New York, 1978.

of magnitude for industrialized countries) if we consider electricity production alone. Thus, the transport sector represents a considerable power generation capability, which, moreover, is increasing as it responds to citizen demands for increased freedom of movement.

The transport sector is an example of a domain that has grown, on a technology basis, without much concern for resource optimization or total cost optimization. The basic need was, and still is, to transport persons or goods as conveniently as possible from one point to another. In the beginning of industrialization, this service was revolutionized by railroad transport. It was followed by the next revolution, the car, first for short-distance transport and distribution, later to dominate all sectors of transport. A greater versatility of services has been provided at great sacrifice of energy economy (Table 4) and the environment.

In most industrialized countries, the social costs, in terms of city planning and emissions, have grown to such proportions that severe changes of strategy for the automobile sector must be taken. Such changes are signaled by the very stringent emission laws passed in California and in the EEC countries. Some of these regulations, such as the more frequent traffic limitations in inner cities, are really aimed at limiting the freedom of movement which has been the reason for the success of the automobile.

Table 5
Emissions from Traffic

Fuel	kg Pollutant/tce fuel				
	SO$_2$	NO$_x$	C$_m$H$_m$	CO	Dust
Diesel	1–4 (0.4)	35 (22)	7 (4)	130 (40)	4 (1.5)
Gasoline	0.1 (0.02)	45 (6)	45 (5)	680 (50)	—

Note: Figures in parentheses are best achievable technology (1990). tce = tonnes coal equivalent.

The pressure of legislation has clearly resulted in big improvements, particularly for gasoline-powered cars through application of the three-way catalyst for reducing carbon monoxide, nitrous oxide, and hydrocarbon emissions (Table 5). Diesel emissions, on the other hand, have not yet experienced the same remarkable progress. This is rather unfortunate, as the diesel engine is thermodynamically superior to the Otto gasoline engine. Despite progress on the emission-reduction side, it appears that the whole transport sector must be looked upon as a system, and solutions considered must provide adequate service at a social cost level that would be acceptable on a sustainable basis. This implies an allocation for continuous growth in this sector due to the expectations of a rise in living standard both within and outside of the industrialized countries. The first steps in this type of clean engineering are the replacement of urban transport heavy diesels with diesels powered by natural gas.

It is interesting to note that the use of alternative fuels such as alcohols has attracted renewed interest, this time mainly as a way of further reducing emissions rather than as a replacement for fossil fuels, although the latter is, of course, still an important aspect in terms of sustainable development. Biomass-based alcohols have an unique advantage over other liquid fuels in that they do not increase global carbon dioxide emissions.

REFERENCES

1. Holappa, L. Metallit suoraan rikasteesta, *Tek. Talous,* September, 2, 28, 1990.
2. Ranta, J., Ollus, M., and Leppänen, A. Issues and problems of flexible paper-production, in Control-90, Helsinki, September 18 to 20, 1990.
3. Ranta, J., Tchijov, I., and Dimitrov, P., *Implementation Strategies of CIM Technologies: Goals and Benefits of Flexibility,* Harvard Business Monogr., 1991.

CHAPTER 8

Materials and Products

When we think about the environmental impact of industrialization, we generally limit ourselves to the problems caused by effluents from industrial processes and frequently forget the biggest pollutant stream constituted by the products themselves, once they have served their useful life. It is only quite recently that one has begun to discuss the fate of used products, generally in relation to the rapidly increasing problem of urban waste.

In a way, this evolution is quite natural. As the environmental concern deepens, we have to move further up the production chain: first, end-of-pipe solutions to primary polluters; later, internal process modifications to reduce emissions and waste and eventually redesign the products to allow maximal recycling of raw materials and minimization of waste production after the products are used. At some point, we may have to redesign the whole system that is providing the service in question if the environmental consequences of the present solution become unbearable. The transport system has been cited as an area in search of new options. In this domain, we are entering a complex chain of new thinking in

167

which the importance of the technical dimension decreases and the social and cultural dimensions grow in importance as we go along.

8.1.   ECODESIGN

The need for new approaches is strongly emphasized by Manzini,[1] who has studied the limits and possibilities of the ecodesign of consumer goods, which, as he points out, constitute the material support for the lifestyles induced by the industrial society. He sees them as the very meeting point of production potential and social imagination for which alternative solutions that show a high degree of sustainability from the point of view of the environment cannot be separated from the extent to and manner in which they are socially and culturally perceived.

According to Manzini, the term "ecodesign" indicates a design activity aimed at connecting what is "technically possible" to what is "ecologically necessary". He differentiates between three levels:

1. Ecological redesign of present products (which, in the light of their whole life cycle, may improve their global efficiency in terms of material and energy consumption, and simplify their disposal or recycling).

2. Design of new single products or services to replace the present ones (which, in terms of performance, may lead to ideas for ecologically more favorable new products or services than the ones presently offered).

3. Suggestions for new environmental scenarios corresponding to new lifestyles (contributing to the creation of new quality principles and to the subsequent modification of the structure of the demand for performance).

He notes that the first alternative implies only technical interventions, requiring no change in life- or consumption styles. The social component is only introduced through market competition, where the product might be given a competitive edge due to its positively perceived "ecological" nature. This concept of "ecolabeling" has been much used, and mis-used, already. The limit of such an alternative is that it confines the development of ecologically acceptable solutions to a system originally conceived and developed outside of any environmental concern.

The design of new single products or services to replace present ones may require at least partial alterations in lifestyle or habits, and these must then be socially acceptable. The advantage is that environmental quality can be given higher priority in the actual design phase, while the limitation is the rather low interest of today's consumer in environmental aspects when it comes to choosing products.

Entirely new environmental scenarios corresponding to new lifestyles requires the highest level of social acceptance which, according to Manzini and others,[30] can only emerge from complex sociocultural innovation dynamics, in which the technical part plays a minor role. Although these alternatives relate to consumer products, they actually have a much wider scope, being applicable to almost all aspects of industrial production. The first two alternatives are the central themes of today's clean production, while the third, despite its potential, has thus far remained at the level of ideological or political manifestations, without sufficient practical substance to attract a wider public.

8.2.   MATERIAL RECYCLING

Raw-material recycling in industry is already a recognized economic activity. Peculiarly enough, the driving force for the reuse of waste materials has, with the exception of scrap metals, been not so much savings in raw material costs as reduction of waste-handling costs. This trend is expected to grow as the full social cost of waste handling is placed on the producer of wastes, as the present trend in the development of legislation indicates.

The economy of waste reduction would be much improved if the design value of the used product could be utilized in addition to its material content. According to Henstock,[2] there is a large difference between the retail cost of a finished product and its eventual value as scrap material.

In some branches of industry, such as used car parts, an established market for certain used components has existed for many years (Table 1).

At present, the high cost of single repair jobs of faulty products prohibits a more widespread reuse of products or even components, as the wage costs in many cases exceed the production costs of mass fabricated items. On the other hand, encouraging examples[3] show that rationally remanufactured goods, if properly designed originally, open

Table 1
Percentage of Reusable Components in Motor Vehicles[1]

Component	% Reusuable components
Engine	25
Axles, differentials, and gears	25
Battery	20
Heater	10
Starter motor, generator	20

up new possibilities for economic benefits in waste reduction and the recycling of materials.

Studies performed by the Fraunhofer Institute in Germany[3] indicate that there already exists a rather well-developed market for remanufactured and reconditioned products such as engines, gearboxes, and generators in the automobile industry, for word- and data-processing equipment in the office machine sector, and for some household devices and electric tools. In the U.S., the practice is even more widespread, due to a larger market involving a subindustry of individual contract remanufacturers, as opposed to the European custom of original equipment manufacturing.

Warnecke et al.[3] have summarized the main advantages achievable through a rational remanufacturing scheme in the following way:

1. Conventional repair procedures with high labor costs lead to uneconomic results since, with regard to processes and supporting aids, they have not even approximately experienced the type of improvements that have taken place in the manufacturing of new parts. In many branches today, an economically profitable repair of products on the market, at a price the customer can afford, is nearly impossible with conventional repair methods (Figure 1). Contrary to this, the manufacture of rationally and batch-produced remanufactured products is an interesting alternative not only with regard to costs, but also as a possible support for market development by the original supplier.

2. The current trend in legislation (increased product and producer liabilities, commitment to delivery of spare parts 7 years after delivery of the product, etc.) will, in the long run, result in repairs or equivalent services after delivery being compulsory. Such a development makes it impossible to run a repair and service department as a separate profit center, turning these departments into subsidized operations representing

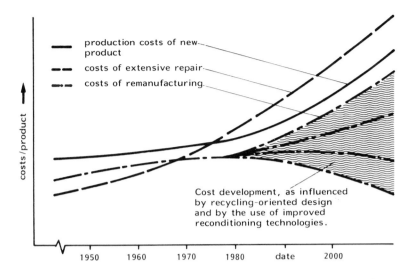

Figure 1. Cost comparison between repair and remanufacturing of products. (From Warnecke, H. J. and Steinhilper, R., 2nd Conf. on Re-manufacturing — Remaking the Future, Massachussetts Institute of Technology, Boston, December 13–14, 1982. With permission.)

a burden to the manufacturer of new products. To combat this, new technologies and organizational forms of after-sales service must take the place of unprofitable repair procedures.

3. Increased costs of raw materials necessitate economic utilization and multiple reuse of the materials. The reconditioning of products into remanufactured goods is a high-value form of recycling. Recycling is generally known as the recovery of raw materials by often energy-intensive separation processes on a large industrial scale. Contrary to this, with the reconditioning of used products, not only the material, but also the function and design value is recovered, so that the efficiency and value level of such a recycling cycle must be rated higher.

Clearly, further development of remanufacturing places additional demands on the designers, who should consider, in addition to the actual functionality of the product, the subsequent remanufacturing operation. Central issues in terms of engineering that must be readdressed are proper choice of materials, development of easy-to-disassemble joints, and procedures for automatized inspection and disassembly procedures. Equally important for economic success will be to find the proper market segments or marketing forms for the remanufactured goods.

Making allowance at the design stage for the introduction of further technical development and modernization into the remanufactured products opens up new challenges for designers and producers.[3]

8.3.   BIODEGRADABLE MATERIALS

Interest in biodegradable materials has increased considerably during the last several years, as they offer potential for reducing the cost of waste handling and the increasing problem of littering, due to faster decomposition of the disposed materials. Furthermore their use is thought to reduce damage caused to fish, birds, and animals by lost fishing equipment and ropes at sea.

On the other hand, the use of biodegradable materials has been criticized on the grounds that some of the materials claimed to be biodegradable actually are not, and that increased use of biodegradable materials may disturb the organization of efficient recycling schemes. Attention must also be paid to trace amounts of catalysts and additives from the biodegradable materials that may eventually have negative effects on the environment.

Several review articles have recently been published on the biodegradability of polymer materials.[4-8] The phenomenon of biodegradability has been studied for decades, with, however, the rather adverse objective of improving the resistance of synthetic materials to microbial attack.[9,10]

8.3.1. Degradation Mechanisms

Degradation of polymers can proceed via different mechanisms. The following classification is due to Swift.[11]

1. *Biodegradation* is an aerobic or anaerobic, enzymatically catalyzed (microbial) process that can take place in a variety of environments (i.e., water, soil, and *in vivo*) and may lead to complete decomposition.

2. *Photodegradation* is a photochemical reaction initiated by electromagnetic radiation such as sunlight, and hence can only occur on the surface of the material. It rarely leads to total decomposition of the material, but may greatly facilitate the consequent microbial decomposition.

3. *Chemical degradation* is a result of reactions caused by the functional groups in the polymer additives and may, as in the previous case, greatly improve biodegradability by microbes.

4. *Environmental erosion* is caused principally by environmental effects such as wind, rain, temperature, or animals. This physical degradation is generally limited to the surface, but again may improve the conditions for microbial attack.

8.3.2. Test Methods

No generally accepted method for classifying materials as biodegradable currently exists, and one must exercise caution in accepting the claims of various biodegradability labels used as marketing tools.

A proper test method involves at least two steps:

1. Incubation of the polymer under defined conditions
2. Monitoring of the decomposition process

Decomposition can be monitored by following the CO_2 evolution, oxygen demand, weight reduction, growth of bacterial mass, or through other physical or visual observation.[5] Furthermore, the test methods can be divided into those suitable for laboratory use and for field tests.[8]

The American Standard Testing and Materials Association (ASTM) G21-70 and G22-76 tests have been used for the evaluation of biodegradability. The methods were originally devised for determining the ability of polymeric material to resist or inhibit microbial or fungal activity. ASTM has established a committee to develop a test procedure specifically adapted for the evaluation of biodegradability (Table 1).[12]

8.3.3. Structural Factors Influencing Biodegradability

The development of polymers has generally focused on strength, durability, lightness, and properties resisting decomposition. There are many reasons for the good resistance against microbial attack (low biodegradability) of polymers. Generally, one might say that the microbes have not had time to adapt and develop specific enzymes for the degradation of synthetic polymers. The large molecular size, small specific surface, and hydrophobicity of polymers also decrease the efficiency of enzymatic degradation processes.[3]

The biological decomposition of a polymer is generally due to hydrolytic or oxidative reactions, and the most important factor with regard to biodegradability is the availability of suitable bonds for these reactions in the polymer chain. The chain must also be sufficiently flexible for an efficient enzyme polymer interaction to take place.

High crystallinity or ordering in the polymer material may slow down enzymatic reactions. The stereospecificity of enzymes is also an important factor in the degradation of substituted polymer materials.[14]

Swift [11] has presented an overview of structural factors influencing biodegradability:

1. -C-C- backbone polymers are generally not biodegradable, with the possible exception of straight-chain oligomers of a molecular weight less than 1000.

2. Polymers having heteroatoms in the backbone are generally more easily biodegradable.

3. Functional groups in the polymer chain influence biodegradability in the following order: esters > ethers > amides > urethanes.

4. Biodegradability is favored by small molecular size, amorphous structure, low strength, and hydrophilicity of the material.

8.3.4. Microbial Polymers

The best known group of bacterially produced polymers are the polyhydroxyalkanoates (PHA), particularly those based on poly-hydroxybutyrates. Many bacteria produce polyhydroxybutyrates as an intracellular energy storage under conditions when carbon is not a limiting factor. The polymer content can reach up to 80% of the dry content of the cellular material.

The commercially most significant biopolymers are the poly-3-hydroxybutyrate (PHB) and poly-3-hydroxyvalerate (PHV) copolymers produced by the bacteria *Alcaligenes eutropus* and marketed by Imperial Chemical Industries under the trade name "Biopol". The production process consists of two steps. In the first step, the cells are cultivated about 60 h until the phosphoric content of the substrate becomes a limiting factor for growth; during the second, 48-h phase, glucose is added and the cells begin to store PHB. If propionic acid is simultaneously added, the PHB-PHV copolymer is obtained, with a PHV content adjustable from 0 to 30%. The polymer beads can be separated from the cellular mass by, e.g., liquid extraction (methanol, methylene chloride) or enzymatically.[15] Biopol is currently manufactured in pilot scale, and the annual production is estimated to reach 5000 tons by 1995.[16]

Using recombination techniques, PHB can be produced by *Eschericia coli* and then separated by a relatively mild heat treatment,[17] but the copolymer has not yet been produced in this way.

8.3.5. Other Natural Polymers

The Battelle Frankfurt Laboratories (Germany) have developed a material that is claimed to contain at least 90% modified starch, the remaining 10% being naturally decomposing additives. The material is transparent and flexible, and is claimed to resist normal conditions well, but decomposes in water or humid soil.[16] It is marketed as a substitute for PVC for packaging.[18]

Warner-Lambert Co. (U.S.) is marketing a biopolymer under the name "Novon" that is claimed to consist almost entirely of modified starch. The material is produced by a melt extrusion process, and is said to have properties intermediate between those of polystyrene and polyethene, with an adjustable solubility in water.[19]

8.3.6. Synthetic Decomposable Polymers

The most important group of synthetic decomposable polymers are the aliphatic polyesters. For example, polycaprolactone (PCL), produced from caprolactone with alcohol as an initiator, has been used in medical and agricultural applications. Poly-α-hydroxyacids such as polyglycolic acid, polylactic acid, and their copolymers have also been used in medical applications due to their biocompatability.[20,21]

Transparent foils for household purposes can be made from the water-soluble polyvinyl alcohols (PVAL). The permittivity of gases of the PVAL foils is very low in dry conditions (relative humidity below 65%) whereas the water solutions of PVAL decompose in nature.[22] Air Products and Hoechst, among others, develop PVAL-based foils for the partial replacement of polyethene.[23]

8.3.7. Mixtures of Decomposable and Nondecomposable Materials

As mentioned earlier, one should not confuse truly biodegradable materials with those made of mixtures of a biodegradable component such as starch and biologically resistant materials such as polyethene, polypropene, or polystyrene, as in an early work by Griffin.[24] A great variety of such materials are presently marketed as being biodegradable, although only the starch component is actually degraded, while the metabolism of the actual synthetic polymer particles is very slow or nonexistent.[25,26] Similar effects can be obtained by the copolymerization of photochemically active groups into the polymer.

8.3.8. Market Outlook

Biodegradable materials are not suited for applications where durability is important, and they should not be used in applications for which the recirculation of the polymeric material can be practically arranged. Major applications are garbage and compost bags, baby diapers and other hygienic products, wrapping foils for food, and containers for detergents, beverages, and some agricultural applications such as planting pots and foils.

The use of biodegradable polymers in 1990 was estimated to be approximately 540,000 tons (about $609 million), out of a total polymer market of about 100 million tons (approximately $100 billion),[28] and the market is thought to be increasing rapidly.[16,23] Of this, the share of garbage bags is currently about 75%, merchandise and grocery bags 14%, and miscellaneous foil products about 5%. Minor applications are six-packs for beverages and baby diapers. Legislative inducement for the development of biodegradable materials is provided by at least the U.S. and Italian governments.[27] Development in the U.S. has been particularly rapid, and 2 years ago some 40 companies were estimated to manufacture products from biodegradable polymers.[23] Information about companies manufacturing biodegradable polymers is presented in Table 2,[27] and a summary of biodegradable polymers is presented in Table 3.[29]

Table 2
Organizations Involved In Degradable Plastics

Organization	Description	Comments
Agri-Tech Industries Champaign, IL	Ethylene acrylic acid copolymer, blended with gelatinized starch	US Dept of Agriculture technology, 45–50% starch loading possible
Air Products & Chemicals Allentown, PA	*Vinex* polyvinyl alcohol	Water-soluble polymers
American Cyanamid Wayne, NJ	*Dexon* polyglycolate	Medical application as suture material
Ampacet Corp Mt Vernon, NY	*Poly-grade* masterbatch	Organometallic additive for photo-degradability
	Poly-grade II	Starch plus prodegradant for bio-degradability, tech-nology from Archer Daniels Midland
	Poly-grade III	Combination of I and II, gving photo- and biodegradation
©Amylum, Aalst. Belgium	Starch-loaded material	Starch will make up 5–7% of PE films; 10–20% PS products; and 6–12% hdPE bottles
Archer Daniels Midland Decatur, IL	*Polyclean* masterbatch	Worldwide licensee of starch technology from Epron Indus-tries, UK
Argonne National Laboratory Argonne, IL	Biodegradable polyactates	Derived from food wastes — possible use in sustained release of fertilizers
Atlantic International Baltimore, MD	*Ecolyte* distributor (see Enviromer Enterprises)	Ketone carbonyl copolymerization
Battelle Institute Frankfurt, West Germany	Starch-based material containing 90% starch	Interest in replacing PVC
Belland Andover, MA	Selectively soluble polymers	Polymer dissolves in water or aqueous base
Dow Chemical Midland, MI	Ethylene-carbon monoxide copolymer Licensing Ecolyte PS technology from Enviromer Enterprises	Used for beverage can retaining rings Photodegradable PS loose-fill packaging

Table 2 (continued)
Organizations Involved In Degradable Plastics

Organization	Description	Comments
Du Pont Wilmington, DE	Ethylene-carbon monoxide copolymer	Beverage can retaining rings
Enviromer Enterprises (Polysar/Ecoplastics) Leominster, MA	*Ecolyte* masterbatches for photodegradable PE, PP, PS	Ketone carbonyl copolymerization
Epron Industries Oldham, UK	*Polyclean* masterbatch	Untreated starch and prodegradant — licensed technology to Archer Daniels Midland
Ferruzzi (Fertec) Italy	Starch-based material containing up to 70% starch	Properties claimed to be similar to PE
ICI (Marlborough Biopol) Cleveland, UK	Copolymer of polyhydroxy butyrate and polyhydroxy valerate (*Biopol*)	First uses in shampoo bottles for Wella of West Germany
Plastigone Technologies Miami, FL	Dual organometallic additives	Used in mulch films
Princeton Polymer Labs Plainsboro, NJ	Aromatic ketones and organometallic salts	Development of products for food packaging
Purdue University West Lafayette, IN	Alloys of cellulose or starch acetates with PE, PP, PS	Shelf stable but readily degradable in landfills or marine environments
Rhône-Poulenc Princeton, NJ	Cerium-based additives	Photodegradation of PE, PP, PS
St Lawrence Starch Co. Mississauga, Ontario	*Ecostar* masterbatch

Ecostarplus masterbatch | Modified starch system for PE film
Ecostar plus Swiss-developed organo-metallic complex |
| Union Carbide Danbury, CT | Ethylene-carbon monoxide copolymer
Caprolactone polyester | Beverage can retaining rings
Containers for tree seedlings |
| University of Massachusetts Amherst, MA | Natural polyesters from bacteria | Material similar to ICI's *Biopol* |
| Warner-Lambert Morris Plains, NJ | Novan `bio-plastic' starch | Starch combined with water to produce a mouldable thermo-plastic |

Rouse, S. ECN Chemscope 55, ECN Environmental Protection Review, July/August 1990, 23–24, 30.

Table 3
Summary of Degradable Polymer Technology, Performation and Market Developments

Factors	Photodegradable products	Starch filled products	PHB/PHBV	Hydrophilic material polymer (starch/water)	Water/alkali soluble polymers
Raw material cost	5 to 10% higher than carrier resins	9 to 15% higher than carrier resins with 6 to 10% starch	1990: = 35 DM/kg 1993: = 17 DM/kg	—	PVAL: 7–10 DM/kg Belland's licensing
Carrier resins	Polyolefins, PS	Polyolefins, PS	—	—	—
Processing	As carrier resins	10 to 30% lower output than carrier resins	More difficult than synthetic polymers (heat stability)	Like thermoplastics, temperature limited at 240°C (patent)	PVAL: more difficult than polyolefins, PS
Degradation Photodegradation	Yes	Possible with additives	Not today	Not today	Water: PVAL
Biodegradation	No	Partially	Yes	Yes	Water/alkali: PVAL, Belland
Rate of degradation	Highly dependent on climatic conditions. Can be controlled	Slow process for starch, difficult to control in landfill	Accelerated process	Accelerated process	
Applications and market potential	Mainly low value commodity packaging — disposable, agricultural market	low value commodity packaging —	High value packaging and specialty applications, initially targeting cosmetic bottles	Development at the beginning, targeted at capsules, injection moulded items, others	PVAL: high value disposable application Belland: specialty application
Development cost	Higher than for commodity resins		Very high	Very high	Belland: high
Market penetration	Slow	Slow	Very slow	Slow, except medical capsules	Belland: slow
Plastics and packaging manufacturers motivation	Prudent and suspicious		Neutral	Neutral	Neutral
Legislative pressure	West Europe: low		USA: high	Italy: very high	

Note: DM = Deutsch Mark

REFERENCES

1. Manzini, E. Limits and possibilities of ecodesign, in Cleaner Production 6.1, Canterbury, September 17 to 20, 1990.
2. Henstock, M. E. *Design for Recycling,* Institute of Metals, Carlton House Terrace, London, 1988.
3. Warnecke, H. J. and Steinhilper, R. Remanufacturing of products in Western-Germany, in 2nd Conf. on Re-manufacturing — Remaking the Future, Massachussetts Institute of Technology, Boston, December 13 to 14, 1982.
4. Klemchuk, P. O. Degradable plastics: a critical review, *Polym. Degrad. Stabil.* 27, 183, 1990.
5. Evans, J. D. and Sidkar, S. K. Biodegradable plastics: an idea whose time has come?, *Chemtech,* 20, 38, 1990.
6. Scott, G. Photo-biodegradable plastics: their role in the protection of the environment, *Polym. Degrad. Stabil.,* 29, 135, 1990.
7. Thauer, A. M. Degradable plastics generate controversy in solid waste issues, *Chem. Eng. News* June, 7, 1990.
8. Abinabhavi, T. M. Balundgi, R. H., and Cassidy, P. E., A review on biodegradable plastics, *Polym. Plast. Technol. Eng.,* 29, 235, 1990.
9. Potts, J. E. Plastics, environmentally degradable, in *Kirk-Othmer Encyclopedia of Chemical Technology,* suppl. vol., 3rd. ed., John Wiley & Sons, New York, 1984, 626.
10. Rodriguez, F. The prospects for biodegradable plastics, *Chemtech,* 11, 409, 1971.
11. Swift, G. Degradability of commodity plastics and specialty polymers, an overview in *Agricultural and Synthetic Polymers: Biodegradability and Utiliztion,* Glass, I. E. and Swift, G., Eds., *Am. Chem. Soc. Symp. Ser.* 433, American Chemical Society, Washington, D.C., 1990.
12. Swift, G. Biodegradation tests and the need for standardization, in Biodegradable Plastics Society, International Symposium on Biodegradable Polymers, Tokyo, October 29–31, 1990, Program and Abstracts, 1990.
13. Narayan, R. Recycling back to nature: Environmentally degradable plastics, in Biodegradable Plastics Society, International Symposium on Biodegradable Polymers, Tokyo, October 29–31, 1990, Program and Abstracts, 1990.
14. Huang, S. J. Effects of structure variation and morphology on the biodegradation and properties of synthetic polymers, in Biodegradable Plastics Society, International Symposium on Biodegradable Polymers, Tokyo, October 29–31, 1990, Program and Abstracts, 1990.
15. Byrom, D. Polymer synthesis by micro-organisms: technology and economics, *Trends Biotechnol.,* 5, 246, 1987.
16. Rouse, S. Biodegradables, friend or foe?, in ECN Chemscope 55, ECN Environmental Protection Review, July/August 1990, 23.
17. Pool, R. In search of the plastic potato, *Science,* 245, 1187, 1989.
18. Anon., In the race for the "greenest" plastics, here are biopolymers, *Mod. Plast. Int.,* 19(9), 7, 1989.
19. Miller, B. All-starch polymer enters biodegradable derby, *Plast. World,* 48(3), 12, 1990.

20. Vert, M. Recent advances in the understanding of the degradation of water-sensitive aliphatic polyesters, in BPS, 1990, 125.

21. Holy, N. L. Vanishing healers, *Chemtech,* 21, 26, 1990.

22. Lorenz, T. Polyvinyl alcohol, a dissolvable "high chem" product — applications for textile sizing, paper finishing, adhesives and others, in Maack Business Services, 1990, 196.

23. Wilder, R. V. "Disappearing" package: pipe dream or savior?, *Mod. Plast. Int.,* 19(9), 74, 1989.

24. Griffin, G. I. L. *Adv. Chem. Ser.,* 134, 159, 1973.

25. Gould, J. M., Gordon, S. H., Dexter, L. B., and Swanson, C. L. Biodegradation of starch-containing plastics, in *Agricultural and Synthetic Polymers: Biodegradability and Utilization,* Glass, I. E. and Swift, G., Eds., *Am. Chem. Soc. Symp. Ser.* 433, American Chemical Society, Washington, D.C., 1990.

26. Tokiwa, Y., Ando, T., Suzuki, T., Takeda, K., Iwamoto, A., and Koyama, M. Development and evaluation of biodegradable plastics containing ester bonds, in Biodegradable Plastics Society, International Symposium on Biodegradable Polymers, Tokyo, October 29–31, 1990, Program and Abstracts, 1990.

27. Rouse, S. (1990), Biodegradables, friend or foe? ECN Chemscope 55, ECN Environmental Protection Review, July/August 1990, 23 - 24, 30.

28. Anon. Degradables market "to double", *Plast. Rubber Weekly,* 1358, 3, 1990.

29. Fuzessery, S. Otherview photo-and biodegradable plastics, in Maack Business Services, 1990, 151.

30. Brown, R. L. Launching the Environmental Revolution in State of the World 1992, WW Norton & Co., New York, 1992.

CHAPTER 9

Policy Questions

9.1. RISK ANALYSIS

The risk for accidental emissions or spills is an often overlooked treatment to the environment, which should and could be taken into account at the design stage of an industrial project. Risk analysis is a developed tool for evaluating (anticipating) damage to installations or human health. However, it has not been fully exploited as a tool for evaluating damage to the environment. Yet, with the sharpening regulations on emissions, we are rapidly approaching a situation in which the major environmental influence from industrial installations might be due to accidental spills or bursts rather than from continuous emissions.

Risk analysis is routinely used within the oil drilling industry to evaluate critical operations. The recent major accidents within the chemical industry such as Flixborough (1974), Seveso (1976), Bhopal (1984), Mexico (1984), and Sandoz (1986) have increased the interest of the authorities in many countries to evaluate the risks of major accidents from land based sources. As a result of this, the so called Seveso Directive has been accepted within the European Community[1] mainly in the interest of protecting human life, but to some extent also to diminish risk to the environment.

9.2.   ENVIRONMENTAL AUDITS

An interesting experiment in the area of preventive environmental protection strategy has been carried out by a research team at the environmental research organization TEM within the University of Lund in Sweden.[2] Essentially, the method could be described as a systematic environmental audit procedure consisting of the following phases.

Informational group meetings — Workshops are held to introduce participating industrial leaders to the general concepts of pollution prevention and waste minimization. Differences between this approach and traditional end-of-the-pipe approaches to pollution control are discussed and successes in similar fields in a number of other countries are described.

On-site meetings — Selected firms in the region are provided with written details of the concepts and procedures to be utilized. Additional discussions are held at the plants to clarify goals and procedures for the program. Details are provided to each firm's representative about the benefits of and procedures for performing waste minimization audits. In the Landskrona experiment the following information was gathered from the companies:

- basic flow diagrams for all the production processes of the firm
- the mass balances of all materials used and produced at the facilities, including all products and emissions to air, water, and soil
- copies of each company's environmental permits were obtained from the local or regional authorities

Plant walk-through — With the background information at hand, the research staff together with the firm's representative performed a plant walk-through to gain a first hand feeling for the production processes and identify possible immediate waste reduction opportunities. Based on this walk-through, more detailed information was requested about the manufacturing processes. A preliminary assessment of the problem was then made by the research team, ranking opportunities for improvement upon which the top priority issues in each firm were discussed with their representatives.

Alternative methods — The research staff then made preliminary assessments of the waste reduction-pollution prevention alternatives for helping the firms to take advantage of the problem solving opportunities identified. These data were transmitted to the company personnel and they were encouraged to analyze the problems in depth and come up with potential solutions or alternatives.

9.3.   REGULATORY INSTRUMENTS

Until rather recently, virtually the only instrument for limiting the environmental damage caused by harmful emissions has been government regulations.

There is no doubt that the national regulations have had a significant impact on changing the course of development, and they have (in a way) acted as a basis for international agreements. As the environmental issues have grown more complex, requiring participation from different sectors of society, (industry and consumers) the inadequacies of a simple regulatory policy have become more and more evident by the rapidly increasing number of regulations (Figure 1).

The environmental problems become increasingly complex and interlinked, and it becomes difficult to devise regulatory documents that take into account all the possible negative interactions, such as shifting emissions between different spheres of legislation, the transfer of pollution from air to water to soil, or interactions between various administrative domains.

As a result, the enforcement mechanisms grow in complexity and, hence, in cost.[3] In terms of cost efficiency, the enforcement measures have been estimated to be from 2 to up to 10 times more expensive than market mechanism related instruments.

The cost factor is important as the expenses for maintaining an environmental policy are rapidly increasing. According to one estimate, they are expected to double during 1990 from current levels of 1.25 to 1.65% of GNP.[4]

However, the most significant drawback of the purely regulatory approach appears to lie in its nature of being a passive and rather static

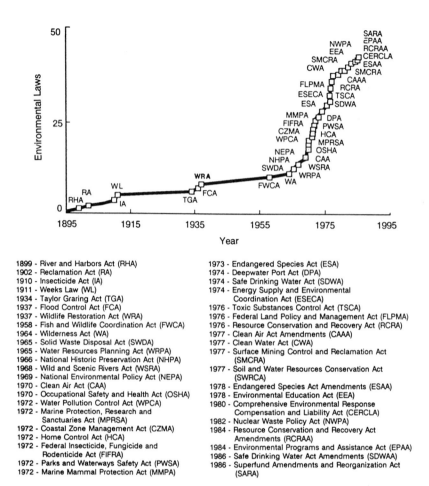

Figure 1.   Increase of number of environmentally related regulations in the U.S. overtime. (From Balzhiser R. W., *Technology and Enviornment,* National Academy of Sciences. Published by National Academy Press, Washington, D.C., 1989. With permission.)

instrument, offering little incentive for technical innovation to go beyond the minimal requirements.

9.4.   MARKET MECHANISMS

As a consequence, more attention must be given to the use of market mechanisms to limit environmental damage, i.e., regulations that would use market mechanisms to guide the development into a direction judged better for the environment.

Market mechanisms now in use, or in preparation, fall into the following categories according to Vernon.[4]

Pollution taxes or charges — This is the well known "polluter pays principle" according to which the charges are adjusted to reflect the true cost to society of the damage inflicted, or to correspond to the corrective actions necessary. Thereby the governments can force the polluters to internalize the social cost resulting from their activities. According to an OECD survey,[4] 14 out of 23 evaluated countries have some kind of instituted taxes on air and water pollution, noise, and potentially harmful products. This approach does assume knowledge or the possibility of monetarization of all the cost incurred by society. In practice, we are far from a situation where these costs can be correctly evaluated. Among the different charges used are:

- effluent emission taxes based on the quantity and/or quality of effluents such as t/a SO_2 or BOD in waste water

- user charges as flat payments for the use of collective pollution treatment facilities, i.e., sewage treatment or solid waste disposal

- product taxes which are charges on particular products to reflect their environmental damage or damage caused by their use, i.e., non-returnable beverage containers, carbon content of fuels to reduce CO_2 emissions

- administrative charges which are usually small scale charges to contribute to the cost of regulatory authorities, for issuing permits. These cannot be considered as being truly related to market mechanisms

- tax differentials where the tax level is varied according to the pollution damage, lower tax on unleaded gasoline vs. leaded on cars equipped with catalyst, etc.

Subsidies — Financial assistance to encourage a switch to less polluting processes or behavior. Investment support for introducing less polluting technology, research assistance, risk loans for demonstration units. In the form of direct grants, loans with reduced interest rates or other favorable conditions or tax allowances for investments resulting in decreased environmental loading.

Market creation — It can be argued that a permit to operate an activity in a certain way also involves a right to pollute according to conditions laid down in the permit. The right then becomes a tradable asset in a way. This thinking has given rise to concepts such as:

- bubbles in which a group of polluters are treated as one single source covered by a giant imaginary bubble. The emissions inside the bubble can be traded freely as long as the total emissions from the bubble do not increase.

- netting, a procedure by which a company can be accorded rights to increase pollution in one sector due to a capacity increase or, similarly by reducing pollution in another

- offset, which allows new pollution sources to be accepted in an area, provided a reduction of the same magnitude is made from an existing source in the same area by an equivalent amount

- banking allows the storage of unused pollution rights for future use, or trade

Although the nomenclature varies in different countries, the principles quoted above are the different approaches used today to employ market mechanisms to enhance pollution abatement activities.

These considerations in particular have been central in efforts to relate the rights and obligations of the industrialized countries towards the developing nations, in relation to CFC and CO_2 questions (Table 1).

9.5.   ENVIRONMENTAL LABELING

Environmental labels are a relatively recent instrument for manipulating the market forces. First introduced in FRG in 1977, the number of products having acquired the right of use of the "Blue Angel" has rapidly increased (Figure 2).

The most important group of products is paint and varnishes (600 items), CFC free spays (177), gas heated furnaces (151), environmentally benign detergents (143), and paper products made from recycled fiber (112). A more advanced version of the ecolabel is the so called environmental product profile concept. In this, a list of environmentally important ingredients consumed in the manufacture of the product are listed on the product package to guide the consumer. The listed values are assembled from the results of an accepted life cycle analysis for the product in question.

Table 1
Waste Emission Taxes in Operation[4]

Country	Purpose	Start year	Coverage	Tax basis	Rate	Revenue raised
Belgium	Incentive to encourage treatment of waste prior to disposal	1981	Industry treating or disposing of waste	Amount of waste composted, incinerated or landfilled (recycled materials are exempt)	0.1–2.6 $/t or cubic meter	Not known
Denmark	Incentive to encourage recycling of waste	1987	Households, industry	Amount of waste generated (wastes classified as harmless (for example straw) are exempt)	6.5 $/t	Not known
Netherlands (surplus manure charge)	Raise revenue for research and pilot projects (some incentive effect expected)	1987 (not yet fully operated)	Farms	Phosphate content of manure produced beyond that permitted to be dumped on land	0.1–0.3 $/t	Not known (not yet fully operated)
U.S.A. (federal and state)	Raise revenue for restoration of hazardous waste disposal sites after closure	1987	Hazardous waste disposal site operators	Amount of chemical waste disposed of	2.3 $/t (federal) up to 72 $/t (state)	Not known

Number of products

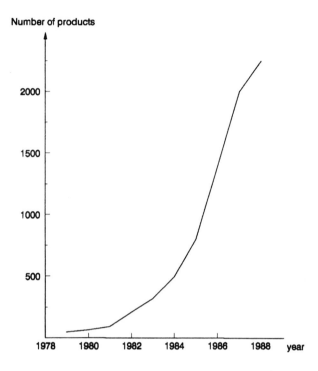

Figure 2. Number of ecolabeled products in Germany, (From Perälä, L., Environmental Labeling of Consumer Goods in Different Countries. Finnish Ministry of Environment Report 67, 1989.)

9.6. THE ECONOMY OF SUSTAINABLE DEVELOPMENT

The general meaning of the expression "sustainable development" is rather self evident; however, the details and implications of it are far from clear. According to Mäler,[5] it could be interpreted as a development scenario in which the future generations are guaranteed a level of income or welfare equal or superior to ours. In this context, the level of income has to be interpreted broadly as the welfare people perceive through material wealth as well as the pleasures offered by the environment they live in. As such a welfare is necessarily subjective, it is impossible to construct an exact measure of the individual welfare adequate for defining the concept of a sustainable economic development.

According to Mäler, the individual welfare is ultimately a result of our possibilities to exploit the productive assets, which in a sustainable development must remain constant over time. Some assets are reduced over time (non-renewable raw materials) others are renewed (forest, air, etc.) and could remain constant over time if they are not exploited faster than they are renewed. If they are underexploited, even a greater heritage can be left to our successors than the one we received.

In addition to natural resources, there are other valuable production assets created by man himself. Eventually, in a sustainable development, the sum of all the assets has to remain constant or increase in value in order to leave future generations better off. The industrialization has resulted in an important redistribution of the natural resources in favor of man-made production assets.

The central question for the future is "Has this redistribution consumed so much of the natural resources that a future welfare is in danger?" The difficulty is that some of the natural assets, clean air, water, and unspoiled nature are difficult to evaluate in exact monetary terms, although they have a very central importance to our well being.

Mäler makes the interesting point that, in general, private ownership has fared better in resource management than temporary or collective ownership as particularly demonstrated by land use in agriculture. Dramatic examples of political mismanagement of resources are provided not only by the ecological disasters in the eastern block countries, but almost in all countries. In particular, within the area of intensified farming, in both the developed and developing world, such examples are legionary.

A fundamental discrepancy lies in the fact that our conventional economic calculations do not include environmental values in a proper fashion. The central problem for defining and inciting a sustainable development is: how can we devise instruments by which individuals, companies, and communities direct their actions towards a common goal of sustainable development?

9.7.   ENVIRONMENTAL EFFECTS

It is important to remember that much of our decision making in specifying acceptable limits for pollution relies on hypothetical extrapolations of laboratory data. The results carried out in laboratories are facts but when these results are transferred to the real world with its multitude of intervening factors, we are making a hypothesis. This is, however, the best we can do as real world experiments often are (depending of the nature of the phenomenon) undesirable or impossible to do.

The use of hypothetical arguments as the basis for restrictive administrative action is acceptable when the menace appears great enough to justify the action. One must, however, be ready to modify the hypothesis if new evidence so demands.

One must also remember that limit values defined on the basis of hypothetical risks are not absolute; the fact that we stay below a certain value does not necessarily mean that the risk for damage is reduced to zero, nor does exceeding the limit by a small amount signify immediate disaster. Despite this vagueness, there is an obvious need to specify exact limits for legal purposes and to enforce them strictly. By the same token, we specify speed limits in traffic and issue fines for nonobedience regardless of the actual absolute risk for the event in question.

A critical area for defining future environmental protection policies, which also has strong bearing on the technical development needs, is the question of critical loads, i.e., the definition of limits of emissions of various pollutants that can be accepted, or should not be exceeded without risk.

REFERENCES

1. Council directive on the major-accident hazards of certain industrial activities 82/501/EEC Official Journal of the European Communities 5.8.1982 p. 230/1-230/18. Council directive amending directive 82/501/EEC on the major-accident hazards of certain industrial activities 87/216/EEC Officiela Journal of the European Communities 28.3.1987 p. 85/36-85/39. Council directive amending directive 82/501/EEC on the major-accident hazards of certain industrial activities 88/610/EEC Official Journal of the European Communities 7.12.1988 p. 336/14-336/18.
2. UNEP Industry and Environment, 12(1), 9, 1989.
3. OECD, Economic Instruments for Environmental Protection Organization for Econmic Cooperation and Development, 1989, 131.
4. Vernon, J. L. Market mechanisms for pollution control: Impacts on the coal industry IEACR/27, August 1990, IEA Coal Research
5. Mäler, K.-G. Proceedings from Ingenjörsforum -90, Helsinki, Finland, 1990.

INDEX